Wilhelm Braune, Edward Bellamy

An atlas of Topographical Anatomy

After Plane Sections of Frozen Bodies

Wilhelm Braune, Edward Bellamy

An atlas of Topographical Anatomy
After Plane Sections of Frozen Bodies

ISBN/EAN: 9783337250133

Printed in Europe, USA, Canada, Australia, Japan

Cover: Foto ©berggeist007 / pixelio.de

More available books at **www.hansebooks.com**

AN ATLAS

OF

TOPOGRAPHICAL ANATOMY

AN ATLAS

OF

TOPOGRAPHICAL ANATOMY

AFTER PLANE SECTIONS OF FROZEN BODIES

BY

WILHELM BRAUNE

PROFESSOR OF ANATOMY IN THE UNIVERSITY OF LEIPZIG

WITH FORTY-SIX WOODCUTS IN THE TEXT

TRANSLATED BY

EDWARD BELLAMY, F.R.C.S.

SENIOR ASSISTANT SURGEON TO THE CHARING CROSS HOSPITAL; LECTURER ON ANATOMY AND TEACHER OF
OPERATIVE SURGERY IN ITS SCHOOL; PROFESSOR OF ANATOMY AS APPLIED TO THE FINE
ARTS IN THE SCIENCE AND ART DEPARTMENT, SOUTH KENSINGTON

PHILADELPHIA

LINDSAY AND BLAKISTON

1877

AUTHOR'S PREFACE

THE accompanying plates, representing plane sections of the human body, are reproduced on a smaller scale from my large atlas, in which the figures are the size of nature; and are intended to assist in extending and increasing the knowledge of the human form, and of the position of the different organs to each other. The necessity for such plates for the clinic has repeatedly been expressed, and especially for military surgery. It has been emphatically stated on all sides that the sections of the thorax and abdomen are indispensable for diagnosis, and that the examination of my plates has been of substantial assistance in judging correctly the direction of a gunshot wound. Consequently when the question arose as to preparing another edition of the large coloured atlas, it seemed advisable at the same time to arrange a smaller one, which could be made accessible to a wider circle of readers. The plates of the large atlas are faithfully reproduced by photography. Some few plates have been omitted as of subordinate interest. The text remains unaltered, with the exception of a few additions.

As each plate has its special text appended, a detailed use of the atlas is rendered possible; and, as circumstances require, recapitulations are introduced for the elucidation of individual plates. In like manner the precise data relating to the structure of the individual subjects are retained and repeated in connection with the plates, in order that the observer, by examining them, may avoid mistakes on the living body.

It will be perceived that the work is not intended to be a text-book of topographical anatomy, but an atlas, which may take its place amongst manuals on the subject, as an illustrated means of assistance.

WILH. BRAUNE.

LEIPSIG;
November, 1874.

TRANSLATOR'S PREFACE

THE great success of Professor Braune's Atlas abroad has induced him to publish a smaller edition of his large work, with photographs of the original plates reduced to half-scale. It has been considered advisable to take advantage of this to reproduce the volume in English.

The immense expense of producing such plates and the persistent dearth of material have, in all probability, been the cause why no original English work on topographical anatomy has as yet been placed within the reach of the generality of students.

It is, I think, generally admitted that there is a want in this country of a good text-book on applied anatomy, and not a mere handbook, but such a work as might take its place with those of Richet, Hyrtl, or Luschka. By means of the sections found in this Atlas the exact position and relations of the structures which must be divided or avoided in the course of an operation are indicated; and the track of a bullet or punctured wound suggested. At the same time they afford an absolutely correct representation of the intimate relations of the viscera of the thorax and abdomen.

I cannot help thinking that the work may be of great value to artists, as demonstrating the exact position of the bones to the muscles and indicating the contours of the body.

I have endeavoured to avoid a slavish translation of the text, and to reproduce the author's meaning in readable English, without interfering more than was absolutely necessary with the original construction of the sentences. I have taken the liberty of omitting some irrelevant matter, such as the repetition of methods of preparation, &c., which would be unnecessary and burdensome to the English student. In the text to Plates XXIX (A, B), XXX and XXXI, finding that the description of the section of the fœtus referred to the author's large work more particularly and was not in any way illustrated by the present series of plates, I omitted it. I trust that Professor Braune will not consider that I have in any way mutilated his text or impaired its utility.

I cannot of course hold myself answerable for the opinions of the author in his surgical and medical comments, but have simply rendered them as I trust he intends them to be understood. I have reduced the measurements to their equivalent English notation (with the exception of the long table on page 155), on the advice of friends, although I consider that it would have been, in some respects, preferable to have retained the metric scale.

I must express my warmest thanks to my friends, Mr. Edmund B. Owen, F.R.C.S., Lecturer on Anatomy in St. Mary's Hospital, and Dr. J. Mitchell Bruce, M.A., Assistant-Physician to Charing Cross Hospital, for their kindness in revising the proof-sheets, and for many valuable hints.

EDWARD BELLAMY.

MARGARET STREET, CAVENDISH SQUARE;
October, 1876.

CONTENTS

Tab. IX.

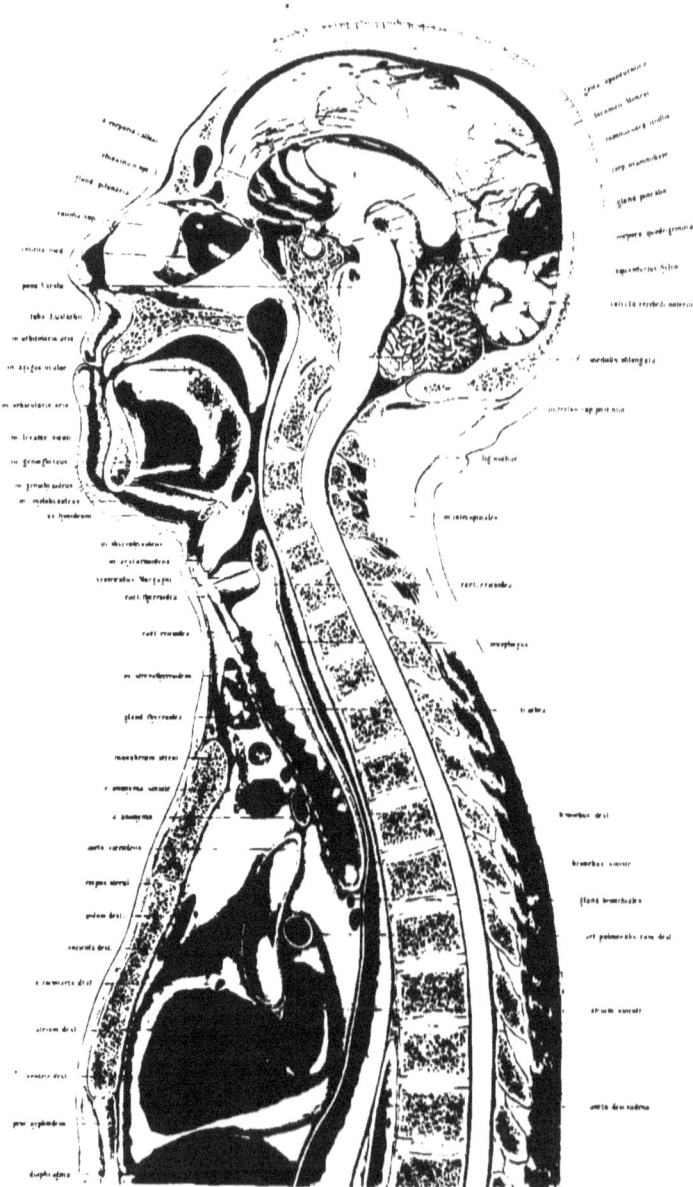

Tab. I n.

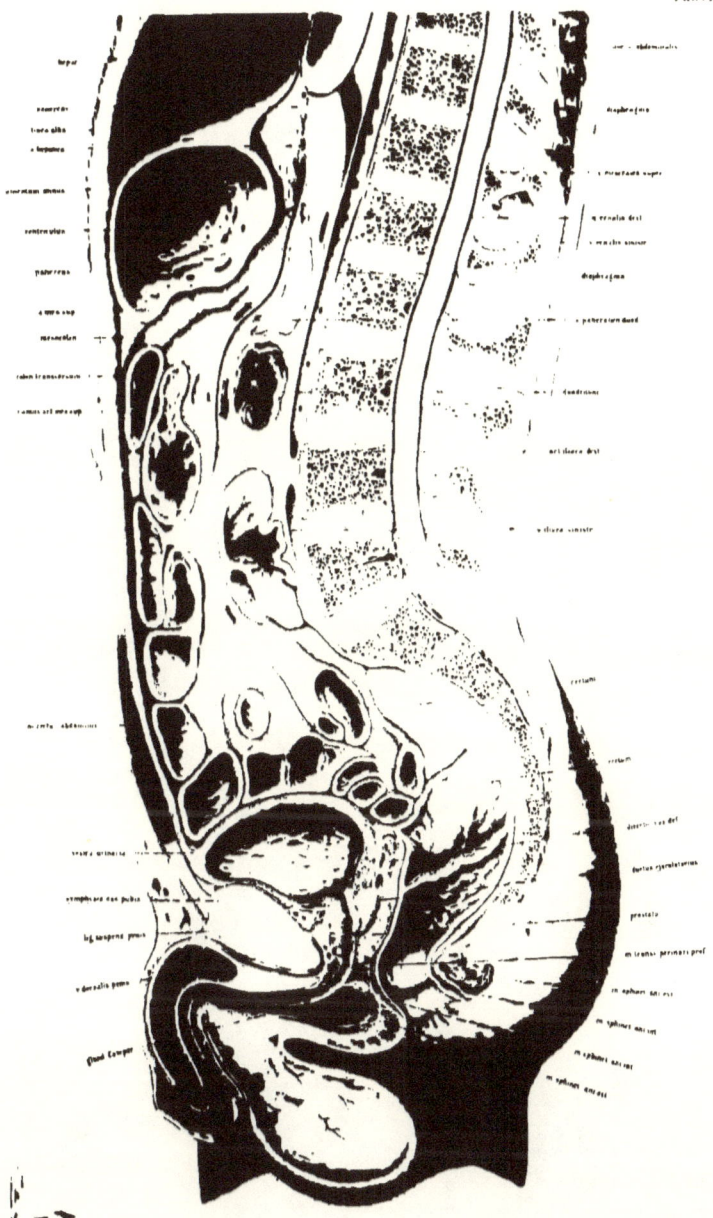

TOPOGRAPHICAL ANATOMY

PLATE I (A, B)

THE accompanying plate was taken from the body of a powerful, well-built, perfectly normal man, aged 21, who had hanged himself. The organs exhibited no pathological irregularities. The body, which was brought in unfrozen, was placed on a horizontal board, without any special support for the head, and it was only by laying it down that provision could be made for the limbs lying as symmetrically as possible with regard to the mesial line. In this position the subject lay untouched in the open air, and at a temperature of about 50° F., for fourteen days. At the end of this time the process of freezing was commenced and completed. The mesial line of the body was next accurately marked out anteriorly and posteriorly with a black line, and the section carefully performed by means of a broad, fine-edged saw, much in the same way as two workmen would saw the trunk of a tree. After cleansing the surface, the right half of the body showed that a most successful section had been made. In the brain the fifth ventricle had been traversed; in the thorax the mediastinum, so that neither of the pleuræ was opened; and in the pelvis the upper third of the urethra. The tracing was then taken from the frozen surface. Where the course of the section had not exactly kept the mesial plane, I improved the preparation subsequently in such places as the nature of the case required. Thus, a thin slice of the cerebellum was removed by means of a razor, and the entire course of the aquæductus Sylvii exposed down to the fourth ventricle, with the penile portion of the urethra and the anus where not opened in the middle line. The plane of

1

the section passed close against the contracted anus, which was opened after
the body had thawed; this accounts for the apparent size of this passage.
In sections which pass through the anus in the frozen condition of the body
the anterior wall lies nearer to the posterior, not, however, so close that
complete apposition is permitted.

It is also to be observed that the details in these plates were worked
out from fresh preparations, in order to produce as useful a result
as possible; due notice will be taken of these details in the proper
places.

With regard to the structures entering into the formation of the skeleton
as seen in the section, the vertebral column holds the chief place. The
section has been so directed that it passes almost through the middle line
of the bodies of the vertebræ; and that the arches, on the other hand,
as is clear in the dorsal region, are divided somewhat to the right of the
middle line.

An examination of the individual portions of the vertebræ shows the
spinal column to be quite normal. No deformity at all was to be found
in the bodies of the vertebræ (as is so frequently the case in aged individuals), but, on the other hand, a great amount of mobility in the parts
was met with, characteristic of a young and actively built person. The
sacrum was devoid of any irregularity, and had a perfect and uniform curve.
That only two portions of the coccyx are to be seen in the plate is
owing to a variation which this part of the skeleton presents, and is not
remarkable.

On examination of the vertebral column in general, its considerable
amount of curvature is first of all worthy of notice.

One would clearly expect that in the horizontal position a flatter curve
would be met with, as the spine, when examined in preparations after the
removal of the thoracic wall and viscera, shows a much flatter arc in the
two halves of the body.

Parow, however, has proved (Virchow's 'Archiv,' Bd. xxxi, p. 108, &c.)
that the removal of the viscera of the thorax causes a great increase in the
flattening of the spinal column. One needs only to compare the method
which was stated by him after the measurement of an isolated vertebra, and

PLATE I 3

is figured *a a O*, Pl. V, fig. 4, with that given by E. Weber ('Mechanik der menschlichen Gehwerkzeuge') and with mine, in order to see at once the great difference.

If the plate before us be compared with that which Pirogoff ('Anatome Topographica,' 1859, fasc. I, A, Tab. 10, 11), made from a body which was also frozen in the horizontal position, and then sectioned, it will be found that the curvatures are nearly exactly the same. Both differ, however, in this respect from Weber's, as they do not show so considerable a concavity in the dorsal region. As Parow found by his observations that the contents of the abdominal cavity, although not on so high a level as those of the thorax, influenced the position of the vertebra, we must look for the cause of this slight difference in Weber's preparation in the previous eventration. Although Weber's proposition for the establishment of the shape of the vertebral column, with its ligaments and discs, is excellent, still it is not thoroughly applicable to all vertebral columns in connection with the soft parts, and must, therefore, be modified according to circumstances.

It would seem now worth while, in the vertebral column before us, to be able to determine what this variation would be in the upright position of the individual, but, unfortunately, the means of doing so are impossible.

If any series of representations of the body frozen in the upright position were given, no advantage would be obtained. It is evident that it is impracticable to keep a body so balanced, and in such equilibrium, as the muscles are capable of doing during life. The trunk always hangs over to one side to such an extent that the spine partly loses its original curvature and takes a semiflexure. It is therefore not to be wondered at that the figure which Pirogoff (*a a O*, Tab. 12) gives, taken from a subject frozen in the upright position, exhibits curves having flatter arcs than it would have had if the drawings had been taken from one frozen in the horizontal position. We should consequently fall into a great error if we conclude on the ground of Pirogoff's plate that in the living individual, whilst in the upright position, the spine has a lesser curvature than when lying down. Parow, indeed, by the help of an instrument (Coordinatenmesser), carried out a number of observations with a view of determining the position of the spinous processes, and so estimated the curvature of the spinal column on the living body.

But valuable as these observations are in an individual case, and however carefully followed out, with a view of showing that each variation of the attitude and balance of the trunk exercises an influence on the position of the vertebræ, it appears to me from the great variation in the forms of the spinous processes, that no absolute rule for the position of the bodies of the vertebra can be adduced, more especially as the exact definition of the promontory still renders special measurements necessary. Therefore I have, apart from this consideration, by comparing Parow's curves with my own plates, estimated the alteration which the spinal column presents in the upright position. An exact determination of the line of gravity of the spinal column in my preparation must likewise be given up. It is not possible to estimate with certainty how this line passes through the individual sections of the vertebræ ; and such definitions can only be undertaken on the living body. If the figure be placed in the upright position, and the head be considered as held forwards, as is the case when balanced on the spine, the excessive convexity in the cervical region becomes somewhat flattened, and a plumb-line hanging from the occipito-atloid articulation would cut approximately the vertebral segments, as the brothers Weber have shown. It passes downwards close behind the promontory and through the line of junction of the heads of the thigh bones, and indeed Parow has by his measurements fallen back on this proposition of Weber's.

Also it is shown by examining the inclination of the pelvis both in my plate and in the one given by Pirogoff, that this is much more considerable than Meyer gives it, and presents nearly the same angle that Weber has determined by his measurements. The line joining the upper border of the symphysis pubis with the promontory of the sacrum makes an angle of 60° with the horizon.

The ligamentous structures belonging to the vertebræ are represented in the plate as accurately as possible. The separate portions also, such as those of the compound ligamentous apparatus of the articulations of the cranium, and those passing down on the anterior and posterior surface of the bodies of the vertebræ, could not be shown in any detail in such a section.

PLATE I   5

However, at the odontoid process of the second cervical vertebra the transverse ligament, with its articulation on the anterior cartilaginous surface opposite the joint fissure between the atlas and odontoid process, is clearly seen, as also are the sharply defined elastic ligamenta subflava. The posterior occipito-atloid ligaments which close in the spinal canal between the occiput, atlas, and axis, have not the elastic quality of the ligamenta flava; they are but slightly distinct from the overlying cellular tissue, and therefore not particularly prominent in the drawing. The section has passed so exactly in the mesial line, that in the neck no muscles are seen except the interspinales, and one in the lumbar region showing through its sheath. In the dorsal region, on the other hand, where the section had passed somewhat to the right side, the tendon-like structure of the multifidus and semispinalis muscles appear. The space between the spinous processes appears in other places filled up with connective tissue, which belongs to the interspinous and supraspinous ligaments derived from the ligamentum nuchæ above. At the inferior end of the spine is seen the posterior sacro-coccygeal ligament, which closes in the end of the spinal canal, and attaches itself to the two portions of the coccyx here shown. The intervertebral discs are represented exactly as they appeared, and their fibrous structure and pulpy centre are clearly shown. It appears that in the most movable parts, such as the cervical and lumbar regions, the discs have an unequable thickness before and behind, whilst those in the dorsal region are of an even thickness. The bodies of the vertebræ in the region of the thorax are of different depths, anteriorly and posteriorly, and consequently influence the curvature of the spine; and it is shown in the region of the neck and loins, which are the most movable, that the intervertebral discs are essentially stronger anteriorly than posteriorly, though the sides of their respective vertebræ are equally deep.

There is nothing peculiar to remark of the sternum and skull; they are sufficiently characterised throughout. The spongy portion is accurately shown in each individual bone of the preparation. Especial care was required to bring each portion of the brain clearly under notice. Sections through fresh brains were used in order that the drawing-in of the parts within the dense contours should be made clear and correct.

Beneath the corpus callosum a good view is obtained of the fornix. It is seen as it passes forward and downward from the splenium, and stopping at the corpus mammillare which lies at the base of the skull. In front of this last lies the infundibulum, which leads to the pituitary body in the sella turcica. Still further forward is a section of the optic chiasma. At the extremity of the fornix is the anterior white commissure. Behind the fornix is the black cleft representing the foramen of Munro, and the inner grey lamina of the optic thalamus with the grey commissure. From the upper white lamina of this some fibres are to be seen passing to the pineal gland, which is in relation inferiorly with the posterior white commissure and the corpora quadrigemina. Beneath the corpora quadrigemina is the aquæductus Sylvii uniting the third and fourth ventricles ; the anterior half of this is covered by the corpora quadrigemina, the posterior half being provided with grey convolutions above from the valve of Vieussens. The floor of the fourth ventricle is formed of grey matter, which is shown to be as a continuation of the grey nucleus of the medulla. This becomes clear from the departure of the posterior fibres of the medulla to the cerebellum.

In the pons Varolii a white band is well seen, the penetrating fibres of the pyramid, whilst those of the olivary body go through between the pons and cerebellum. Behind the pons is seen a portion of the nucleus of the olivary body cut through. Between the several portions of the brain which are not directly in apposition, the sites of the great subarachnoid spaces are seen. One, for instance, between the anterior (here upper) border of the pons and the corpus mammillare, and a second between the cerebellum, the medulla, and the commencement of the spinal cord ; a third between the posterior part of the corpus callosum and the cerebellum. The investing arachnoid, which, springing across from one portion of the brain to another, so forms this space, cannot be reproduced in the plate on account of its excessive fineness. Excepting the artery of the corpus callosum, which passes upwards over the genu, all the vessels depicted are veins.

The superior longitudinal sinus is laid open for almost its entire extent. The inferior longitudinal sinus on the lower border of the falx is only to be distinguished by the blood seen through its walls. Beneath the splenium the

PLATE I 7

vena Galeni magna passes upwards in order to empty into the straight sinus, of which only a small portion is met with at its junction with the lateral, whilst the thyroid plexuses of the third and fourth ventricles are very evident and clearly represented in the plate. The dura mater, which in the cavity of the skull lies close down upon the bone and on the foramen magnum, and is connected with the external periosteum, leaves the bony walls in the spinal canal and approaches the cord. At the commencement of the cauda equina at the lumbar vertebra the cord can (in the plate) be no longer distinguished from the dura mater.

It will be observed that a portion of the septum narium has been removed. This has resulted from its deflection towards the left side. It was not caused by a polypus. I amplified the defect somewhat in order to bring the relation of the mucous membrane to the septum narium and the two upper turbinated bones clearly into view. Behind the septum is seen the inferior opening of the Eustachian tube. It follows from the relation of the parts, that instruments which are introduced into the tube must be passed along the floor of the nares in order to preserve the necessary direction. The plate shows that an examination of the opening of the Eustachian tube by means of the laryngoscope, would be materially facilitated by drawing the velum forward and upward. The relation of the uvula to the glands and muscular tissue is evident. The thickness of the velum must be borne in mind in the operation of staphyloraphy. One is inclined to underrate its thickness, and thus to experience difficulty in freshening the edges of the cleft.

Mouth.—Before the freezing of the subject the contents of the stomach had ascended into the œsophagus, and partly filled up the cavity of the mouth. After removal of the frozen mass its tube could be represented in the plate.

It can be seen also in the present preparation that the tongue is formed like a muscular pestle, which can thrust hither and thither the contents of the cavity of the mouth. The relation between the tongue, hyoid bone, and larynx is clearly shown. If the surgeon desires to reach the larynx easily, he only requires to draw the tongue out of the open mouth, and can then move the epiglottis and with it the larynx upwards and forwards. The parts of the hyoid bone and the neighbouring organs, which are here shown, are similar to those represented in Pirogoff's plate, and as it was not taken from a person who had died by hanging, they may be regarded as normal. The larynx is evenly divided in the mesial plane, and offers no peculiarities for consideration. The sections of the cricoid and thyroid cartilages, and the ventricle of Morgagni between them, are shown, and, on account of the apposition of the vocal cords, the ventricle appears only as a cleft. The muscles to be noticed in this section are, on the posterior wall of the larynx, the transverse section of the arytenoideus, anteriorly, between the cricoid and thyroid cartilages, some fibres of the crico-thyroid lying close in the mesial line, and above a portion of the thyro-hyoid.

The ligaments shown are the glosso-epiglottic, the middle thyro-hyoid, and further down the middle crico-thyroid.

The section of the neck is so closely in the mesial plane that no vessels are seen, except a vein above the manubrium sterni, a communicating branch uniting two subcutaneous veins of .the neck. It lies enclosed between two laminæ of fascia, which arise from the splitting of the anterior lamina of the cervical fascia. Behind this lies the cut edge of the sterno-thyroid muscle. Between this muscle and the trachea is the section of the middle portion of the thyroid body which is perfectly normal in its relations. The plate shows the direction taken by the knife in tracheotomy, and the importance of keeping the incision exactly in the middle line of the neck.

The absence of arteries in the middle line, as is almost uniformly the case, shows that there is less apprehension of danger in the middle line from hæmorrhage than laterally. The thyroidea ima artery is the only one which would be met with in such a plane, and this, according to Neubauer, is found in one in every ten bodies. Since this vessel takes its origin in almost all cases from the innominate its distribution must be looked for somewhat towards the right of the middle line. As the trachea lies further distant from the surface of the body as it descends, the operation of tracheotomy is easier of performance the nearer the surgeon approaches the larynx, consequently, unless there are contra-indications, it should be performed above the thyroid body. It must be recollected that this

PLATE I 9

gland should be drawn upwards by a blunt instrument in order to freely expose the upper rings of the trachea, a proceeding unattended with difficulty owing to the mobility of the organ. Should the operation be performed below the thyroid body there is a considerable depth of tissue to get through before reaching the trachea, and, moreover, great attention must be paid to the position of the vessels of the neck. The position of these trunks is not so constant that any general rule for their distance from the upper edge of the sternum can be given.

The trachea, which in this preparation divides into the two bronchi opposite the fourth dorsal vertebra, has tolerably the same relations, as shown by Luschka (' Brustorgane,' Tübingen, 1857). It appears, however, from sections on other bodies that there is no constant point of division, and different authors make different statements on this matter. Henle (' Anatomie,' 1866, Bd. ii, p. 26 A), describes it as opposite the fifth dorsal vertebra. Pirogoff in his plate (Fasciculus I A, tab. 14), gives it as high as the third.

Thorax.—The slight depth of the thorax is striking, and yet one can convince oneself, both from measurements on the living body and also from Pirogoff's plates, that there is in this case no abnormality. The mediastinum was so exactly divided by the section that neither pleural sac was opened; whilst of the lungs, nothing is seen but a small strip of the right, which, covered by pleura, is shown behind the body of the sternum. In Pirogoff's plate (Fasc. I A, tab. 10, 44), no lung is to be seen, by reason of the considerable breadth of the mediastinum. The heart was so divided that only a flat piece of the arch of the aorta remained in the right half of the body, whilst the root of the pulmonary artery was removed with the left side, its right branch being cut through The superior and inferior venæ cavæ are not seen at all, they lie deeply, and empty themselves above and below into the right auricle, so that their point of entrance cannot be clearly made out. If, in the plate, a line be drawn from the anterior border of the septum auriculorum outwards and downwards, the situation of these deeply lying vessels will be indicated. The large cavity in front of and below the aorta belongs to the right auricle, the larger portion of which remained on the right half of the body. Its cavity extends upwards toward the right auricular appendix, of which, as is clear by the plate, only a small portion reached across to the left half of the body posteriorly towards the vertebræ, and somewhat behind the left auricle. A large portion of the tricuspid valve has been removed in the section.

Only a small portion of the left auricle is left, and this is seen lying behind the right auricle, and between it and the spinal column. About two thirds of it were removed with the left half of the body. The two openings into it correspond to the entrance of the pulmonary veins. That portion of the auricular septum containing the foramen ovale is removed, and only a small portion of the right ventricle is noticed.

Here the heart was cut obliquely near its upper surface, and there-

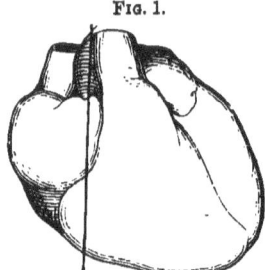

FIG. 1.

fore its muscular tissue and fatty layer appear remarkably clearly. There is a considerable amount of fat on the heart. The muscular struc-ture of the heart and valves, however, shows no irregularity. The relation of the pericardium is clearly shown. The accompanying woodcut ex-plains the position of the heart with regard to the mesial line as found in the present case, from whence result the rules for its percussion. It will be noticed that the relations agree exactly with those given by Luschka (loc. cit., tab. iii).

The entire length of the œsophagus is not distinctly shown by the median section, as in certain places the tube diverged considerably from the middle line. In this preparation, however, on account of the contents of the stomach having regurgitated into it, it was so distended that the plane section met it throughout its course.

Abdomen.—It can be seen from the form of the abdominal walls, that there is no sinking-in of the parietes, but, although the intestines were moderately distended, the short distance of the umbilicus from the lumbar vertebræ is very remarkable. The depth of the abdomen in the mesial line is, indeed, very variable, and is generally represented far too great.

But it is to be expressly noticed here that the condition of parts seen in the present drawing is not precisely the same as in the living body, since in

PLATE I 11

the dead subject the lungs are in the position of fullest expiration, and the diaphragm reaches its highest level; and the relation of the intestines with it, the distribution of the blood, and the arching forward of the abdomen, are somewhat altered. Therefore, with reference to the living body, the distance of the vertebral column from the abdominal walls must be considered as somewhat greater, although not so much so as one is accustomed to suppose. From this relation of the parietes to the vertebræ, the possibility of the ready compression of the abdominal aorta may be inferred. Compression becomes the easier the thinner the individual and the less full the intestines. Further, it is evident that the individual should lie in such a position that the lumbar vertebræ be bowed as much forward as possible; and as the aorta bifurcates on the fourth lumbar vertebra, the pressure should be brought to bear directly on the navel.

Intestines.—The position of the intestines in the middle line should be compared repeatedly with other sections on bodies of the same size. It appeared that a similar figure continually obtained, and that, with exception of some of the coils of intestine, the stomach, duodenum, transverse colon, iliac flexure, and rectum, when in an equal state of distension, lay pretty much in the same position. In one case the stomach was found in such an empty and contracted condition that it was at first entirely overlooked, and when it was found the little finger could be scarcely pushed into its cavity. On examining the abdomen it appears (and more so than in other regions) that the change in the volume of individual organs as well as their mobility may be considerable without other parts having essentially to suffer thereby. For fat and cellular tissue so completely surround the viscera that no empty spaces are left, and thus freedom of movement and compression are permitted.

The section of the liver passes through the left lobe near the lobulus Spigelii.

The pancreas is cut through near its head, where the superior mesenteric vein approaches the liver. The other part of it, which is directed from the head of the gland to the middle line along the lower horizontal portion of the duodenum (the so-called lesser pancreas), lies behind the mesenteric vein, so that it looks as if the vein passed through the pancreas itself.

The relations of the peritoneum are represented in the plate as they were met with after the thawing of the preparation; only, for the sake of clearness, half the fat of the greater bag of the peritoneum has been taken away and the layers thereof shown somewhat diagrammatically.

A vertical section in the middle line is not the most favorable for showing the mutual disposition of the reflexions of the peritoneum; an oblique one taken outwards from the foramen of Winslow, through the root of the mesentery to the iliac flexure, would much better answer the purpose. Therefore, in the accompanying woodcut I have given a diagrammatic representation, which will at least make clear the relation of the lesser bag to the other portions of the peritoneum. The individual layers of which the transverse meso-colon is composed, are not represented in this drawing as they cannot be prepared in the full-sized body, and their diagrammatic representation would only complicate the drawing.

On the relations of the rectum there is nothing further to add. The distance of the peritoneal sac from the anus, which is here about three inches, is to me noticed, as is also the position of the so-called valves of the rectum. Since the rectum in its ascending portion courses over towards the left half of the body, there is only a flat section of it to be seen; in this respect my plate differs from those of Henle and Kohlrausch.

The representation of the bladder also differs from that given by the above-mentioned authors, it was, however, accurately drawn from the preparation. The bladder was completely full of frozen urine, and consequently there was no sinking-in of its upper wall, as is represented in several of Pirogoff's plates. I injected the bladder with tallow as soon after death as possible, partly through the urethra and partly through the ureter, both in the vertical and horizontal position, in order to compare the form and situation of that viscus. A section in the mesial plane in each case showed the same conditions as in the plate, and, with reference to the flattening of the upper wall, no essential difference was found whether the body was upright or lying down.

The position of the entrance of the urethra corresponds with Henle's and Kohlrausch's description, though no absolute similarity need be expected. Langer ('Med. Jahrb. Wien.,' 1862, 3 Heft) has shown that many consider-

PLATE I 13

Fig. 2.

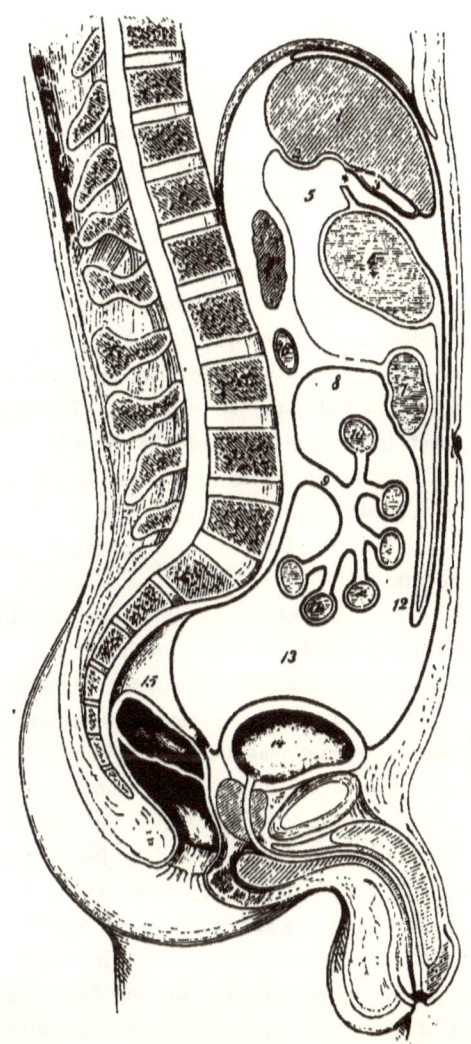

1. Liver cut obliquely.	6. Pancreas.	12. Great omentum.
2. Lobulus Spigelii.	7. Transverse colon.	13. Cavity of peritoneum.
3. Gall-bladder.	8. Transverse meso-colon.	14. Bladder.
4. Stomach.	9. Mesentery.	15. Rectum.
*. Foramen of Winslow.	10. Jejunum.	16. Duodenum.
5. Lesser omentum.	11. Ileum.	

able variations obtain as regards this matter. Especial care was expended
on that envelope of the bladder which forms the porta vesicæ of Retzius, as
this is not very clearly shown in Henle and Kohlrausch. It is shown that
from the termination of the posterior wall of the sheath of the rectus (the so-
called fold of Douglas) two laminæ of fascia take their origin, and then pass
down close to one another between the rectus and the peritoneum. If the
bladder be only moderately distended, as in this case, they however confine
a space in front of the peritoneum, which is taken possession of by the
bladder as it rises upwards during distension. The anterior lamina passes
downwards as a thin covering upon the rectus abdominis and lines the
space between the bladder and the symphysis pubis ; the posterior lamina
passes across behind the urachus on to the bladder, in order to invest it,
and to join the prostatic capsule and pelvic fascia. The internal vesical
sphincter is clearly seen in the plate, but, on the other hand, the external
sphincter is not completely brought into view. The limits of the prostate
gland are clearly defined, also the parts lying in front of the urethra are
accurately represented. In most cases the muscular fibres and gland tissue
are not exactly made out.

In front of the prostate is the middle pubo-prostatic ligament with
the numerous veins which form the plexus venosus of Santorini. Beneath
it is some muscular tissue which has not been completely analysed.
It was represented as it stood, and, after Henle, is comprehended under
the name of deep transverse perineal muscle ; it, moreover, corresponds
with Muller's so-called constrictor of the membranous urethra. The triangular ligament of the urethra (Colles), which lies on the ligamentum
arcuatum, close beneath the symphysis, and is incorporated with the
deep transverse perinei, does not appear very clearly defined in this
section. The white portions on the anterior border of the above-mentioned
muscular mass are to be referred to this. Vertical sections in an
antero-posterior direction are not adapted for the demonstration of the
pelvic fasciæ and muscles ; those made across the axis of the body afford
better results.

The dorsal vein of the penis and the suspensory ligament are well
shown.

PLATE I   15

The curvature of the urethra differs somewhat from that which Kohlrausch describes as normal, but the condition here represented must also be regarded as such, since it presents no pathological irregularities nor are there any in neighbouring organs. It must be therefore assumed, as follows from the plate of Pirogoff and Jarjarvay, that this urethral curvature which offers in the normal condition frequent variations can only be generally defined. Moreover, the ease with which instruments can be introduced into the bladder merely by their own weight proves that it is less a question of giving the catheter a definite curvature, than of knowing of the hindrances which might oppose its introduction. The projection in the prostatic portion of the urethra corresponds to the prostatic sinus near the colliculus seminalis, which lies in the section with the ejaculatory duct.

The relations and structure of the glans and corpus cavernosum are well shown, so also is the fossa navicularis. The other dilatations and contractions of the urethra which are regular in the normal body cannot be defined. In order to obtain a clear idea of these, casts must be made from soft specimens as Langer has done, as sections of hard preparations are not of much value. The position of Cowper's glands, which lie so deeply below the urethral muscles, will explain why the inflammation and enlargement, which are frequently found on section to have affected them, are so little regarded during life; a considerable amount of swelling must occur in order to afford any perceptible tumour.

If the plate be examined with regard to perineal operations, such as lithotomy, one is astonished at the narrowness of the space between the upper portion of the urethra and the rectum. It must be remarked, however, that in the present instance it is peculiarly exaggerated, as the rectum was full of fæces.

The importance of the rule is evident that before the operation of lithotomy be undertaken the rectum be cleared of all fæcal matter, in order that it be out of the reach of the knife. That the space is thereby substantially enlarged is manifest from Kohlrausch's plate, which is drawn from a greatly distended rectum.

It is further seen from the relations before us, that it is quite practicable to preserve the capsule of the prostate. By dilating the membranous and prostatic portions of the urethra more room is obtained for entering the bladder, as well as for the removal of large calculi. By the preservation of the posterior part of the prostate with its capsule, dangerous urinary infiltration is obviated. With regard to the high operation of lithotomy above the symphysis there is nothing to remark. The plate also shows that the bladder must be fully distended in order that that part of it which is not covered by peritoneum may be raised sufficiently above the level of the pubic symphysis.

Tab. II A.

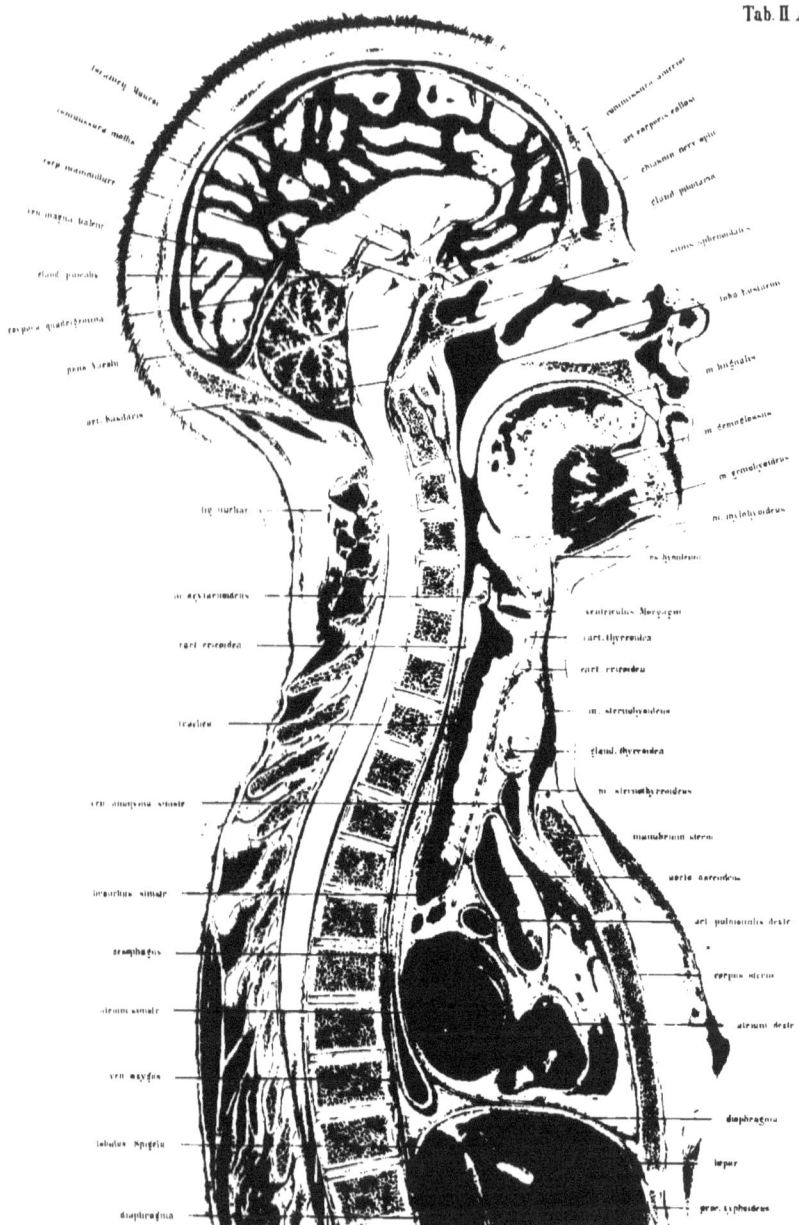

Tab.II B.

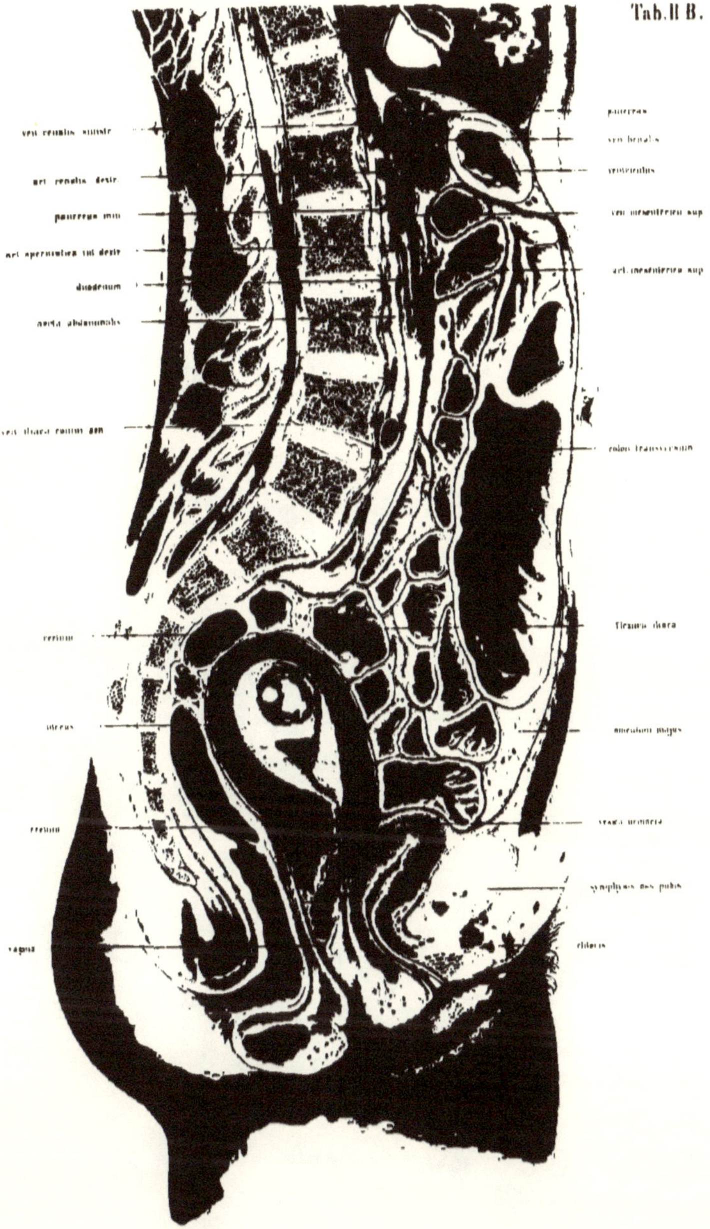

veni renalis sinistr
art renalis dextr
pancreas min
art spermatica int dextr
duodenum
aorta abdominalis

vena iliaca comm dextr

rectum

uterus

rectum

vagina

pancreas
ven renalis
ventriculus
ven mesenterica sup
art mesenterica sup

colon transversum

flexura iliaca

omentum majus

vesica urinaria

symphysis oss pubis

clitoris

PLATE II (A, B)

THIS section was made on the body of a finely formed woman (twenty-five years of age), which was brought into the dissecting room immediately after death by hanging. The arteries were injected with paint, the body laid on the back and frozen, and the details of the section carried out as in the last case.

The uterus was found to contain a foetus of, probably, the eighth week. All the organs were normal. The stomach and intestines were tolerably empty; the transverse colon was moderately distended with flatus, and the rectum with faeces. The bladder was contracted, and as no urine had flowed from it during the transport of the body, it was probably empty at the time of death.

The section was carried from below upwards, chiefly in order to divide the pelvis in the middle line, and was, on the whole, very successfully directed. The articulation of the symphysis was opened, and so also were the urethra and lowest part of the rectum.

On the other hand, the uterus, which lay somewhat on the left side, was cut through in its right half, yet so near the middle line that it was necessary to remove a thin slice only in order to show the canal of the cervix throughout its extent. The spinal canal was opened throughout, and very near to its middle line.

It will be noticed, from the appearance of the dorsal portion of the cord, that at the lower part of the thorax the vertebræ are cut to the right of the middle line, and from the appearance of the great vessels of the abdomen, that the section passes through the diaphragm between the caval and the aortic apertures. The inferior cava is entirely removed with the right half of the body, and a transverse section only of the left common iliac vein is

3

seen; the abdominal aorta, on the other hand, is completely shown, with the right common iliac artery divided.

In the thorax the saw has passed exactly in the middle plane; neither lung is seen and neither pleural sac. As regards the tongue, a small lamina only had to be removed to expose its mesial plane. The cerebrum was not cut exactly in the middle line, so that about one tenth of an inch of the dura mater had to be removed in order to expose the longitudinal sinus and to accurately halve the brain, which had been in the meanwhile hardened with spirit.

Before I enter upon the chief points of importance in this plate or describe the pelvic viscerá, I shall point out the general relations of the parts, commencing with the vertebræ.

The spinal column shows a very beautiful curve, which contrasts favorably with that in Plate I. On account of the slight bending backwards of the head the cervical vertebræ do not project so far forwards, and the dorsal spine does not curve backwards so considerably, but passes more gradually into the convexity of the lumbar curve.

If a line be drawn parallel with the long axis of the body, commencing in the region of the occipito-atloid articulation, and then passing through the posterior border of the odontoid process of the second cervical vertebra, it would touch the last cervical and first dorsal vertebra (in Plate I it touches the three lower cervical), and pass down close behind the promontory. The line passing nearly through these points is, according to Weber, the line of gravity.

The inclination of the pelvis is 58° (less than that of the male in Plate I, which is 60°).

The slight projection of the promontory is characteristic of the female spine, as opposed to that of the male, and so also is the more abrupt direction of the symphysis pubis. It is evident from this circumstance that the conditions are more favorable for the expulsion of the child, which thus glides the more easily downwards on to the promontory from the more abrupt surface of the symphysis. It is repeatedly contested that the axis of the symphysis (by which is understood the direction of the greatest length of the joint) is more abrupt in the female than in the

PLATE II 19

male, and from this an impediment to parturition has been sought. I am not able to declare whether in this particular a constant difference exists between the male and female pelvis. From a series of sections on frozen bodies I have, however, found this relation over and over again, as this and the first plate show, and I might therefore direct the attention of gynæcologists to this point, for I am unable as yet to give any decided opinion upon it.

The conjugate diameter is very large, 4·8 inches. The pelvis, on the whole, is wide, but is not otherwise abnormal. There is not much to remark as regards the head; the individual parts are the same as in Plate I. It is fortunate that the mouth was firmly closed, as the two incisor teeth shut upon one another like the blades of a pair of scissors. The tongue completely filled up the mouth. In a transverse section of the tongue a shallow furrow is generally noticed at its back, which passes from before backwards, a narrow space being left between the tongue and hard palate; hence it must be assumed that the middle line of the tongue was not in this case exactly in the line of section. The œsophagus, in which was some undigested food, admits of delineation throughout its entire extent, but on account of the shading, it is not satisfactorily represented in its original position; against the third dorsal vertebra the shading is not intense enough to show its deep excavation. At the level of the sixth and seventh dorsal vertebræ, on account of the small piece cut off, more of the œsophagus lies on the right half of the body, and consequently its course forms a flat S-curve in the frontal plane.

In front of the trachea lies in section a considerably developed thyroid body, which causes a slight bulging forwards of the neck. Beneath this lies the left innominate vein, and close to it are the remains of the thymus gland; behind the vein is the ascending aorta with a section of the innominate artery. The course of the innominate artery with regard to the trachea is of considerable surgical importance. An incision made in the mesial line of the neck between the thyroid body and the upper border of the sternum would reach the vessel as it lies on the trachea. Ligature of this vessel has hitherto not been successful, owing to shortness of the trunk (from one inch to an inch and three fifths). It is not to be wondered at that the conditions for the formation of a firm coagulum are here unfavorable. It must also be borne in mind that the incision made to search for the vessel is, like that made in tracheotomy, below the thyroid body, and that at the lower end of the wound the left innominate vein may be met with. The trachea, which when extended lies on the anterior surface of the œsophagus, divides into its two bronchi in front of the fourth dorsal vertebra, as is shown in the section represented in Plate I.

I was much surprised by the apparent shortness presented by the trachea in the section of a frozen body made with the head depressed, and by its becoming very considerably extended, when at the commencement of thawing I reinstated the head in its normal position. It is owing to this extensibility of the trachea—due solely to the elastic tissue between the cartilaginous rings—that positions of extensive flexion and extension of the head can be taken up without thereby causing dislocation of the roots of the lungs. Were the trachea a uniformly solid tube it must follow that at each flexion of the head it would be pushed dangerously upon the root of the lung and left auricle, whilst on each abrupt jerking back of the head the thoracic viscera would be dislocated upwards by the sudden drag. Measurements which I have made show that the amount of extensibility of the trachea during flexion and extension of the head is about one inch, and that there is no considerable folding or pinching-up of tissue in its inner wall. This peculiar condition also accounts for the wide gaping of all transverse wounds of the trachea during extension of the head.

Of still further practical importance, particularly with relation to the performance of tracheotomy, is the variation in the relative position of the trachea and the anterior surface of the neck in the different positions of the head. During extreme extension of the head the trachea is brought considerably nearer the surface of the neck, and is consequently more accessible; moreover, the field for operation is much more extensive than when the chin is in the usual position of depression. The section given by Pirogoff (l. A, 14, 1) is remarkably instructive on this point. Again, with the extension and advancement of the trachea, the arch of the aorta and the

PLATE II   21

innominate artery are drawn somewhat higher, and in this way the latter vessel is rendered more accessible for ligature.

As regards the heart, its left auricle was distended, owing to the injection having entered it from the lungs, thus the appearance presented by these parts is normal. The oval-shaped section of the distended left auricle is seen close to the œsophagus, before the more triangular opening in the right auricle. A small portion of the right ventricle is opened by the section. From both auricles the corresponding ventricles can be seen through the auriculo-ventricular openings; those parts, after careful cleansing, are shown in their hardened condition. In the left auricle is seen the entrance of the pulmonary veins, in the right the coronary sinus. The sinus, with the valves of Thebesius, are shown in the lowest part of the triangular section of the right auricle. A portion of the valvular apparatus can be seen in the divided arch of the aorta; behind the vessel lies the right branch of the pulmonary artery. A small portion of the right auricular appendage which was left in the left half of the body (also agreeing with the section in Plate I), was removed, so that a considerable space is left in front of the aorta inside the pericardium.

If the section of the thoracic cavity be compared with that of the young man (Plate I) it will at once be observed that the upper border of the manubrium of the sternum is half the depth of a vertebra higher in the male, and about ⅛th of an inch further from the spine than in the female. In the female the upper border of the sternum corresponds with the space between the second and third dorsal vertebræ. The greater capacity of the male thorax is also demonstrated from the fact of the diaphragm reaching to the level of the fibro-cartilage between the ninth and tenth dorsal vertebræ, whereas in the female its highest point corresponds with the upper border of the ninth, and is consequently the depth of an entire vertebra higher. We have to deal here with a well-proportioned though greatly developed female, but as the two subjects were of the same age it will be of great advantage to compare them. It appears that the position of the several parts of the heart in both is nearly similar as regards the mesial line. (In both cases the auricles and a small portion of the right ventricle appear in the section.)

Nothing is seen of the lungs in young persons in such a preparation in consequence of the presence of the thymus gland; in the condition of expiration their anterior edges never reach the middle line, consequently a median vertical section does not expose lung tissue. In old persons, in consequence of the dwindling away of this organ and of the slight capability of contraction of the lungs, they meet one another after death anteriorly; and, moreover, the right lung frequently overlaps the left half of the body.

On account of the slight distension of the intestines the cavity of the abdomen showed but little prominence, but not, however, an actual in-drawing of the abdominal walls as one observes in sections of bodies which have become emaciated from sickness. In this case, from the amount of fat beneath the skin and in the abdomen it is plain that the individual was well nourished. Also in this particular the circumstances closely resemble those of Plate I, although there is a considerable difference with regard to the depth of the abdominal cavity. In consequence of the greater distension of the stomach and intestines in the male subject, which is manifest from the greater extent of the section through the intestines, and that in the female the arteries were injected and the gravid uterus pushed a portion of the small intestine upwards, the distance of the abdominal wall from the vertebræ at the level of the twelfth dorsal vertebra, in this plate, amounts to, nevertheless, 2 inches less, whilst in the region of the umbilicus the depth of the abdominal cavity is much the same in each, viz. about 3·5 inches. It is, moreover, to be borne in mind that in the male spine the concavity of the dorsal region begins lower down, is more decided than in the female, and further, that the bladder in this instance is empty and in the other tolerably full.

The section has so fallen through the abdomen that the diaphragm has been met with between the œsophageal and caval openings more towards the right side of the spine, so that the abdominal aorta is not divided as in Plate I, but remains intact on the upper surface. In order to make the artery more clear for the drawing, only a small layer of cellular tissue was removed so as to render distinct its plastic appearance. At its inferior extremity is the divided right common iliac artery; nothing is to be seen

PLATE II   23

of the inferior cava (which remains in the right half of the body) but a small portion close to the left common iliac vein. In like manner (as in Plate I) the trunk of the superior mesenteric vein is divided at the point where it, after receiving the splenic vein, courses over to the right side, opposite the pancreas, and passes to the liver as the portal vein. In front of the lower end of this vein is the superior mesenteric artery.

The pancreas, though not so broad as in Plate I, has a similar position at the level of the first lumbar vertebra. The superior mesenteric vein passes through the (lesser) pancreas throughout its extent. The duodenum, which was tolerably empty and flattened by the injected vessels, appears as a narrow cleft in front of the second or third lumbar vertebræ, at the inferior end of the lesser pancreas. In Plate. I, in consequence perhaps of the greater development of the lesser pancreas, it lies somewhat deeper.

A small piece of the lobulus Spigelii of the liver, which is covered by peritoneum, is seen remaining in the left half of the body. The complicated arrangement of the peritoneum in this region can be understood by consulting Plate XV, which represents a transverse section at the level of the eleventh dorsal vertebra, and thus accidentally corresponds to the section which separates both plates.

The stomach was empty and contracted, but the transverse colon, which was considerably distended with gas, hung down like a sling, and was therefore divided to a greater length. There is no peculiarity to be noticed in the small intestine. A portion of the ileum is pushed up out of the pelvis by the uterus, and therefore the lumina of the intestines fill up the abdominal cavity higher than in Plate I. We must here consider the relations of the rectum more attentively. It was evenly distended with frozen fæcal matter, and was of great calibre. The anus is directed backwards as in the upright position, a direction dependent on the inclination of the pelvis; but in the sitting position, when the equilibrium of the trunk is maintained by the tuberosities of the ischium, the symphysis is raised so considerably that the conjugate diameter is nearly horizontal, and the anus takes a direction directly downwards. Above its lowest curve, at the level of the coccyx, is a transverse fold, which is the commencement of the valves of the rectum of Kohlrausch. Higher up the rectum gradually passes over towards the left side, and afterwards it crosses the middle line again by a sharper curve to fall a second time into the plane of section. From the transversely divided lumen of bowel, which lies in front of the third and fourth pieces of the sacrum, the rectum appears again more in the middle line, and following the curvature of that bone, terminates in the iliac flexure. The rectum thus forms a double **S** curve; one portion lying in the antero-posterior plane of the body, the other in the transverse. These bendings serve to support the sphincter-apparatus during the pressure of the fæcal matter, so that at the time of defecation a resistance is afforded which would not exist were the direction of the rectum vertical. It will be observed also that the name *rectum*, which has been applied to this portion of the intestine, is incorrect; it originated from the old representations which were made from undistended intestine and soft preparations.

In front of the rectum, between it and the contracted bladder, is the gravid uterus. Considerable interest is claimed for this section, from the fact of the womb being in a state of gestation corresponding with the end of the second month.

I am unable to say how it happens that the body of the uterus is so sharply bent against its neck and turned backwards, for its tissues are absolutely normal, and according to the statement of Holst ('Beiträge zur Geburtskunde,' 1 H., Tübingen, 1868, p. 162) at this period of pregnancy anteflexion rather than retroflexion would be expected. I can only with difficulty accept the proposition that the uterus during life had some other position originally, and that directly after death, when the body was placed on its back, it sank down from its own weight. At the same time it must be admitted that the space between the uterus and rectum was previously occupied by small intestine, and yet we cannot imagine that they slid upwards in order to make room for the body of the uterus. The subject presented throughout firm tissues and strong muscles, and there were no signs of a previous pregnancy. The relations of the intestines are normal; no coils lie between the uterus and rectum, or uterus and bladder.

PLATE II 25

The deep situation of the external orifice of the uterus, from which a firm plug of mucus projects, corresponds with early pregnancy. Later on the uterus rises up out of the pelvis and draws the vaginal portion up with it, so that the external os takes a higher position.

The uterus itself inclined somewhat towards the left side, so that the plane of section passed obliquely through its long axis, and only a small portion of it was removed with the right half of the body.

The hinder lip of the cervix appeared as if it had slipped away, and wanted only a thin slice more to be cut off in order to expose the canal of the cervix throughout its length.

The bag of the amnion was untouched, and the umbilical vesicle was clearly evident. I have removed from the wall of the uterus successive layers so that. the individual parts of the ovum may be seen distinctly.

On the inner side of the muscular tissue of the uterus can be seen the decidua vera, consisting of uterine follicles, cellular tissue, and blood-vessels. The round openings of the follicles could be easily seen with the naked eye on the inner and upper surfaces. Above, commencing in the anterior wall of the uterus, the decidual layer is extremely thin, but it gradually increases on the posterior surface, and in the neighbourhood of the internal uterine orifice it is still thicker. Corresponding to the thinnest spots, at about the middle of the fundus, the decidua reflexa is shown as a fold over the triangular clot of blood. It is one of the thin envelopes of the ovum, and is most external. It is formed from the chorion lævis and the decidua reflexa, and upon it are found remains of epithelium, connective tissue, and rudimentary tufts.

From the position of the effused blood (which is accurately represented) a slender, whitish line runs backwards and upwards, dividing the chorion frondosum from the decidua vera. The portion of the chorion which is shown in the plate contains only tufts and vessels; it indicates the place of formation of the placenta. In this neighbourhood the umbilical cord is already discernible as it runs deeply downwards.

Inside the chorion was a viscid fluid, in which floated the sac of the amnion, the vitelline duct, and umbilical vesicle. Distinct membranes

4

between the chorion and amnion were not made out in the fluid. The
embryo shows the usual curvature of the trunk with the head bent
forwards. Its length, from the coccyx to the head as it lay in its original
position was about four fifths of an inch, and when stretched out it was
about one inch and a fifth.

The cranium was so enveloped by its coverings that the division of the
brain could not be seen clearly through them. The nose was small, but
already formed. The lateral parts of the oral cavity (the cheeks and lips)
were already so developed that the mouth appeared as a circumscribed
fissure. The upper and fore arms were flexed and separated; the hands
were discernible and the lower extremities were in a proportionate stage of
development. These conditions correspond with the development of an
embryo described by Erdl ('Die Entwickelung des Menschen und Hühnchens
im Eie,' Leipzig, 1845, taf. iii, 6, iv, 18, ix, 3 and 4).

The umbilical vesicle is represented too tense and large. It lay on the
closed amnion as a flaccid bag, looking like a membranous disc, of the size
of a lentil, or about one fifth of an inch in diameter.

The vagina, divided through its anterior and posterior rugæ, appears as
a narrow fissure, and is continued upwards behind the posterior lip of the
external os.

The right sacral ligament of the uterus is seen in the bundles of fibrous
tissue here divided. The fibres do not admit of being clearly discerned
from the muscular tissue of the uterus, but show themselves merely as a
transversely divided bundle, the continuation of which in the fold of
Douglas grasped the rectum on both sides and extended to the sacrum. It
cannot be accurately defined where the vagina ends and the uterus begins.
The muscular fibres of both organs lay so close together, and were so inter-
laced, that they are represented as being in continuity. It is clear from the
plate that the peritoneum stretches further down behind the uterus than in
front, and that it covers a small portion of the wall of the vagina. A thin
process of fascia is united to this by lax cellular tissue, which permits of a
shifting of the rectum and vagina in their mutual distension. I have found
the connexion between the bladder and cervix uteri so arranged that the
possibility of considerable distension of the bladder, and of the anterior

PLATE II   27

wall of the vagina, and a rising of the uterus would be permitted. The ascent of the base of the bladder is not possible if it were, as is stated by Courty, adherent to the cervix (Courty, 'Maladies de l'uterus,' Paris, 1866, page 11).

The clitoris is shown with its right crus divided. Behind it, and in front of the urethra, lie the divided blood-vessels of the bulb of the vestibule. The urethra opened of itself after the thawing of the preparation, and is shown in the dilated condition. In the tissue in front of and behind the urethra was found (by means of the microscope) a layer of striped muscular fibre which forms the sphincter of the urethra. I was not able to obtain a well-made female subject for section, nor did I succeed in getting a body affected with well-marked anomalies in as regards the position of the uterus. It is, however, of great practical interest to compare the plane sections of such a body with the plate; I therefore give a series of reduced copies from Legendre and Pirogoff, with the idea of making this work as complete as possible. The bladder in fig. 1 contained, according to Legendre, nearly one pint of urine. The peritoneum at its reflexion from the bladder was two inches from the symphysis, and its distance behind the uterus from the perineum was 2·08 inches. The conjugate diameter was 4·2 inches; that of the outlet was 2·4 inches.

Although a decidedly atrophic uterus is here represented, the plate must be accepted, as I sought in vain for a better specimen. The distended bladder has lifted the peritoneum some way from the symphysis, and has pushed the uterus downwards and backwards. The form of the bladder is not that usually found in young persons. The spindle shape which is so characteristic in children persists for some long time, whereas the rounder form is met with in old persons. It is not improbable that Legendre had removed the viscera from the subject the preparation was made from, which would account for no sections of intestines being seen excepting the rectum, as well as for abdominal walls being cut off shorter than the spinal column. Should this have been the case, the form he has given to the bladder is explained; a bladder lying freely is distended in a different manner from one which lies in the closed abdominal cavity, as can be easily proved by experiment. On the other hand,

the level of the reflexion of the peritoneum above the symphysis agrees with
that given in the young subject which Legendre figures. In young persons
only can we reckon that in distension of the bladder so much space would
be gained that in extracting a calculus above the symphysis a sufficiently
large vertical incision can be made without wounding the peritoneum ; in

Fig. 1.

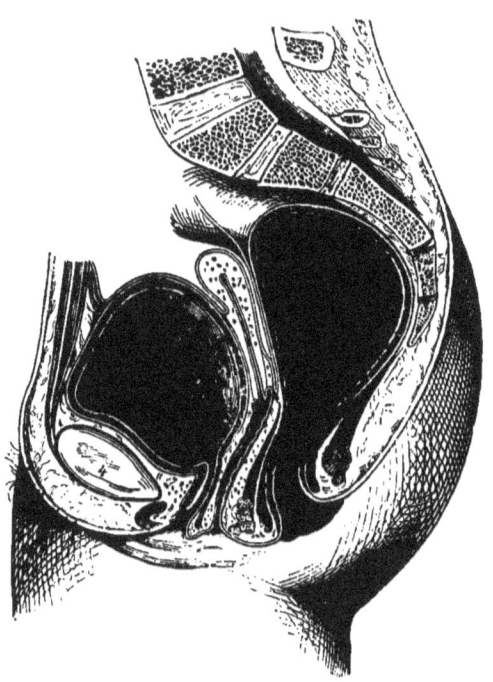

Pelvis of virgin, æt. 18.   Legendre, ' Anat. Homalographique Paris,' 1868, pl. xvii.
         1. Uterus.   2. Bladder.   3. Rectum.   4. Symphysis.

old persons the space is very limited in a transverse direction.   Between
the uterus and the rectum no coils of intestine are seen.

The following plate from Pirogoff merely shows the bladder and urethra
fully distended in order to demonstrate the anatomical relations of the high
operation and the vestibular incision for stone.

PLATE II   29

A thorough distension of the bladder left the peritoneum an inch and a half from the symphysis, and by the traction on the anterior wall of the

Fig. 2.

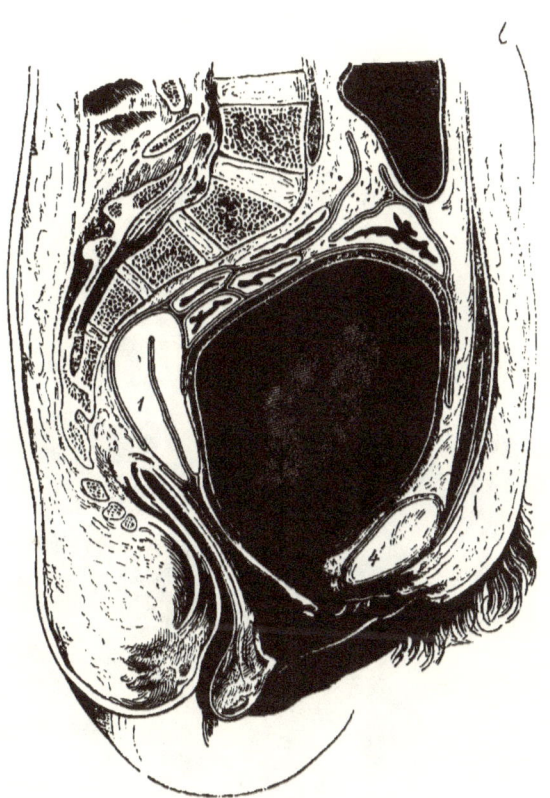

Female pelvis; bladder and urethra distended.   Pirogoff, viii, A, 32, fig. 20.
1. Uterus.   2. Bladder.   3. Rectum.   4. Symphysis.

vagina the uterus is drawn upwards and backwards.   The conjugate axis is 4·08 inches.   The rectum is empty and contracted.

The extent to which the position of the uterus varies with distension and evacuation of the bladder is well shown in figs. 2 and 3.   The section in fig. 3 has not gone exactly through the mesial plane, and thus, although the uterus is bisected, the anus and the urethra have escaped division. In fig. 2 we have exactly the opposite conditions, namely, an empty bladder

Fig. 3.

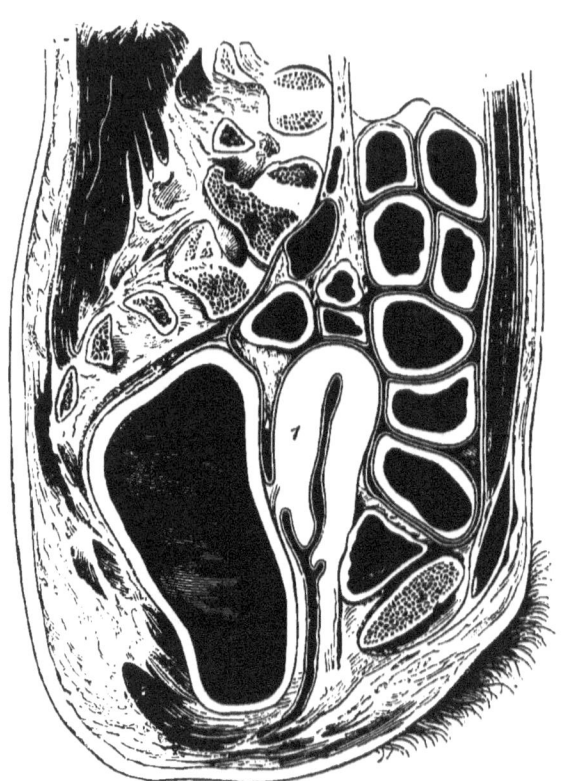

Female pelvis, æt. 35; normal; bladder empty, rectum distended. Pirogoff, iii, A, 21, fig. 3.
1. Uterus.　　2. Bladder.　　3. Rectum.　　4. Symphysis.

and distended rectum; consequently the relations of the uterus and vagina are altered. Whereas in fig. 2 the uterus follows the axis of the vagina, in this instance it forms an obtuse angle with it, without, however, being

PLATE II   31

anteverted.   No coils of intestine lie between the uterus and rectum.   The conjugate diameter is 4·41 inches.

The uterus in fig. 4 with all its connections, was normal, and lay between

FIG. 4.

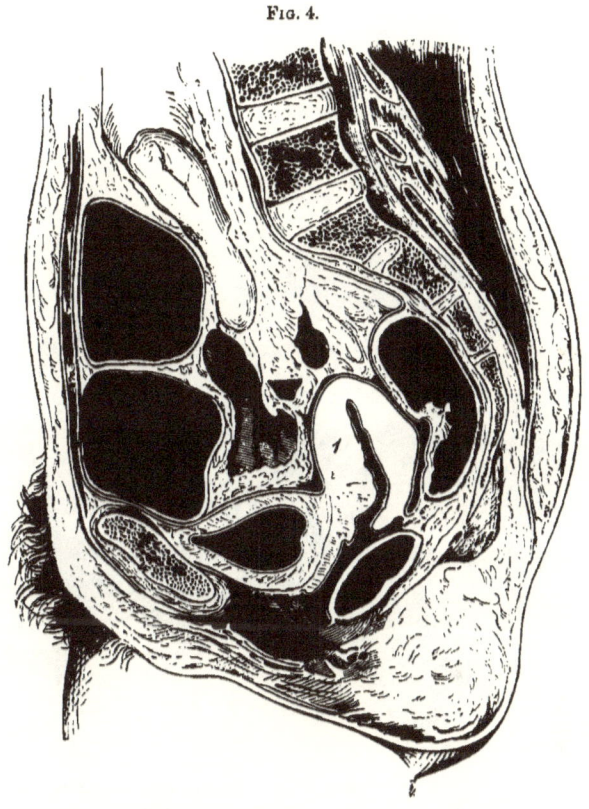

Pelvis of a female of middle age; multipara; normal.   Pirogoff, iii, A, 22, fig. 1.

1. Uterus.   2. Bladder.   3. Rectum.   4. Symphysis.

the moderately distended bladder and rectum; nor do coils of intestine lie behind the uterus in this section.   It will be noticed, therefore, that in the different degrees of distension of the bladder and rectum the uterus is

always in the middle line between these viscera, whilst its position varies
with its volume.

The uterus in this figure lies considerably deeper than in the foregoing
ones.   The conjugate diameter is 4·2 inches.

FIG. 5.

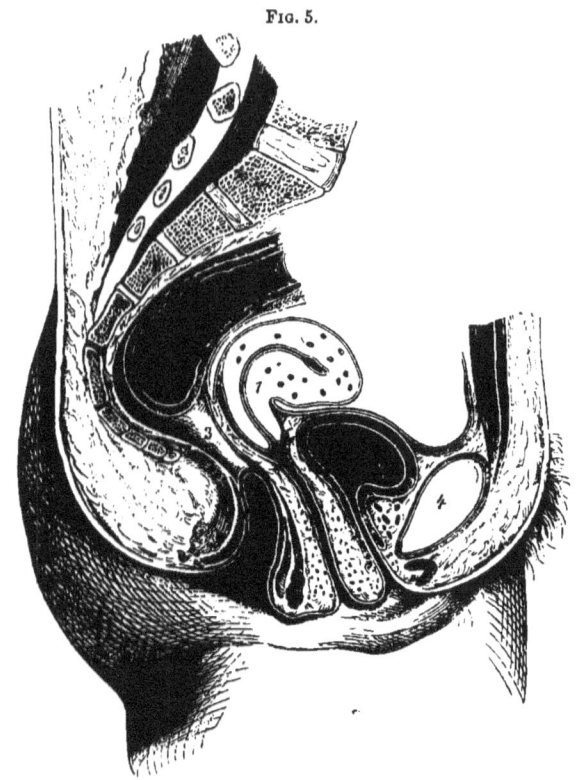

Female pelvis, æt. 30; multipara; anteflexion of uterus.   Legendre, xviii.
1. Uterus.   2. Bladder.   3. Rectum.   4. Symphysis.

Fig. 5 (from Legendre), from a section of a frozen body, made after
the removal of the viscera, shows well-marked anteflexion of the uterus.

The angle between the body and neck of the uterus impinges upon the
rectum.   The walls of the uterus appear throughout of uniform thickness,
and the rectum and bladder are but slightly encroached upon.

PLATE II 33

The bladder may be so compressed in the middle line that it assumes an hour-glass form, one portion of it still retaining the urine after the other has been emptied by the catheter. In such conditions the catheter would have to be passed into the further cavity, so that all the urine might be drawn off.

The anterior lip of the os is continued into the anterior wall of the vagina without a clearly defined border, whilst the hinder is strongly prominent and has a length of one inch. The cavity of the vagina contains the canal of the cervix. The vagina itself is 3 inches in length, whilst that in fig. 4 was only 1·5 inch, and the long extended one in fig. 2, 2·8 inches. In like manner the distance of the peritoneum on the posterior wall of the vagina from the perineum is increased, being 3·24 inches; in fig. 1 it is 2·08 inches. The conjugate diameter is large, being 4·28 inches. The ante-flexed position of the uterus is shown by Schultze to be the normal one in young persons when the bladder is empty. The uterus would, following the contracting bladder, lie upon it, and from traction exercised by the utero-vesical ligament of Courty extend the base of the bladder backwards (cf. Volkmann, 'Sammlung Klinisches Vorträge,' No. 50). There is no question that during the variations of the forms of the rectum and bladder, according to the amount of their contents, a space remains in the pelvis near the uterus, which must either be temporarily filled with small intestine, or render necessary a larger amount of variability in the shape of the uterus itself. If we exclude with Claudius and Hennig the possibility of a filling-up of Douglas's pouch by the small intestine, in the case of the bladder and rectum being empty, the difficulty of representing the topography of the uterus would be enormous, as is evident to every experienced anatomist. We have the choice only, either with Henke to show the uterus set up at a fixed angle with the vagina surrounded by small intestine, or with Schultze to represent it bent over on the bladder. However important it may be to determine these relations accurately, I do not think that it can be done at present; in any case I should not follow Schultze's statement completely. The extension of the base of the bladder does not appear to me in Schultze's plate to be correct, still less so does the assumption of a forcing of the same by means of a ligament, as

Courty describes.   The lax cellular tissue which lies between the uterus and
the base of the bladder, and in which a large number of thin-walled veins
run, cannot be regarded as a ligament in the usual sense of the word, and

Fig. 6.

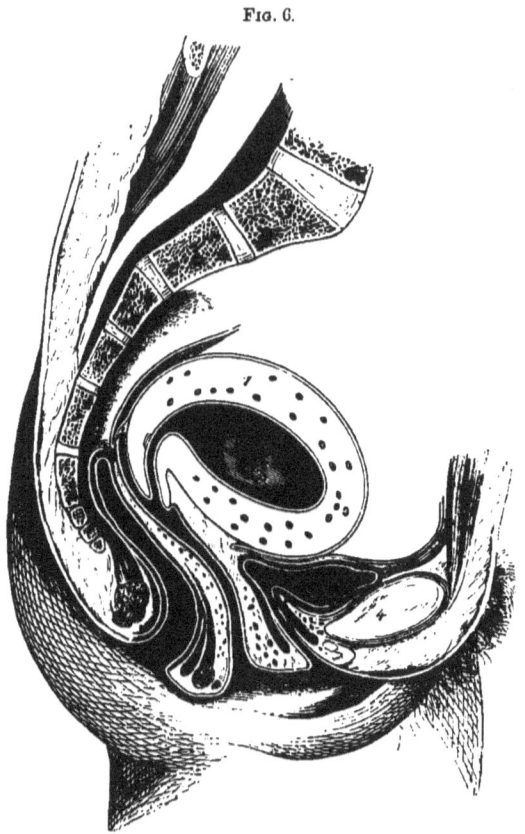

Female, æt. 35, after childbirth; ante-flexion.   Legendre, xix.
1. Uterus.   2. Bladder.   3. Rectum.   4. Symphysis.

is not shown as such in my plates.   It would be necessary to obtain a
series of bodies of young females in order to study the variations of the
position of the uterus in well-hardened preparations.

PLATE II   35

The woman (fig. 6) died immediately after childbirth; the anteflexion also was recent, brought on by the weight of the heavy body of the uterus, which had a capacity of about two fluid ounces.

The flexion is so considerable that the body and neck of the uterus are almost at a right angle with each other, and on the anterior wall a distinct fold is formed. The posterior wall of the uterus rests on the rectum, and presses on its lumen, &c. The vagina is distended and measures 3·6 inches. The distance of the peritoneum on the posterior wall of the uterus from the perineum is 3·8 inches. These figures consequently considerably exceed those in the preceding case.

The position of the fundus with regard to the firmly-compressed bladder is to be remarked here, so also the position of that portion of the peritoneum which lies between the uterus and bladder on the anterior wall of the vagina. In the normal condition of the uterus its end lies nearest the peritoneum, whereas it is here in the middle. It is further to be noticed that the strong attachment of the posterior portion of the base of the bladder with the neck of the uterus (which as already mentioned is admitted by Courty) is not present in this preparation, otherwise the bladder and urethra lying close down on the uterus would be dragged upwards. Nevertheless, the lax cellular tissue and fascia between these viscera is not so capable of distension that variations in the position of the uterus could continue without any influence on the bladder. We notice here that a large piece of the posterior portion of the base of the bladder has been drawn upwards, a condition which would interfere with the action of the vesical sphincter, and consequently cause an incontinence of urine. The conjugate diameter is very large, 5 inches, and exceeds that in Pl. II.

The ante-flexed position of the uterus is very well represented by Legendre, as is also to be found in Pirogoff's 'Atlas,' and in Rüdinger's 'Topogr. Chirurgische Anatomie,' i, II Abtheilung, taf. ii.

The female from which his · plate was made died shortly after instrumental labour, and was soon given over to Rüdinger, who froze it and obtained a good section. The thick-walled uterus had dragged the small intestine upwards and lay with its fundus high above the symphysis. Here also the base of the bladder had not followed the traction exerted by the ante-flexed uterus, so that the statement of Courty is not borne out in this case.

Fig. 7 shows a retro-flexion of the uterus. The thick hyperæmic uterus, which has been opened at one spot only (1), contains masses of coagulated · blood, and shows retro-flexion as well as a lateral deviation. The body of

FIG. 7.

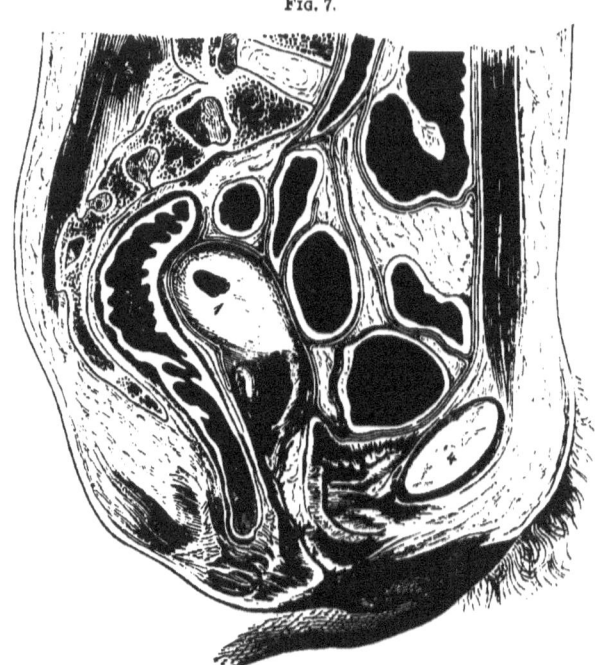

Female pelvis; retro-flexión of uterus. Pirogoff, iii, A, 31. 14.
1. Uterus.   2. Bladder.   3. Rectum.   4. Symphysis.

the uterus lay more in the left half of the pelvis, and the neck, which was divided throughout its length, kept its original position. Such a condition of the uterus would cause a pressure on the rectum which would be increased to complete compression were the retro-flexion more extended.

Therefore, if the retro-flexion be very considerable a stoppage of fæces may be expected, but ante-flexion, as the preceding figure shows, is able to produce a similar result. The conjugate diameter was 4·4 inches.

PLATE II   37

If, with a view of forming any conclusions, the figures here given bo compared, it is at once noticed that the statement of Claudius ("Bericht über die Naturforscherversammlung zu Giessen," 1865, 'Zeitschrift für Rationelle Medizin,' iii Reihe, 23 B., p. 244), according to which the normal uterus is not so movable and not so enclosed on all sides by coils of small intestine as is generally represented, is quite borne out. The uterus lies so much more between the bladder and the rectum, that Douglas's pouch contains no small intestine. Corresponding with this is the gravid uterus in Pl. II, A, B, in closer apposition with the hinder wall of the pelvis. However, as I have mentioned above, I am not able to express myself precisely on this matter.

It would be important to institute further investigations on pregnant bodies to find out whether the sharp bending seen in Pl. II, A, B, is post mortem or not; at all events, the fact of lying the body on its back cannot alone be the reason of it. If the uterus were ante-verted, and directly after death sank from its own weight into the position of retro-flexion, we must expect that during life there may have been similar variations when a supine position is taken up repeatedly and for a long time at once. On the other hand, I cannot assent to the statement of Claudius, when he affirms that the bladder by its filling and emptying communicates no movement to the uterus. I have already mentioned that one can demonstrate that an influence is exerted by the full and empty bladder upon the site of the uterus on the dead body, also on the living subject the variations of the position of the uterus from micturition can be made out. A cup-like sinking-in of the upper wall is shown, moreover, in the empty bladder only, when the subjects are not very fresh and the urine has been voided immediately before death. The round and empty bladder with its firmly contracted thickened walls (Plate II) shows the normal relaxations of this viscus when it has been emptied by its own power of contraction.

PLATE III

PIROGOFF, in his works which we have frequently cited, has carried out his sections of the human body in three directions, and upon that attempts to establish its configuration. It might be expected, as a matter of course, that such a proceeding could be applicable in geometrical solids, but it is not sufficient in organised bodies, even if one divided such a body into as many thin laminæ as possible. In organised bodies oblique sections in individual places show far more than purely sagittal, frontal, and transverse sections.*

This is more particularly the case with the eye. Supposing it be desired to give not merely the relation of the globe, but to include the structures in its immediate neighbourhood, we must aim at bringing into view the connections of the eye with the optic nerve as far as possible up to the brain.

It is of only secondary interest to determine in which horizontal plane behind the orbit the individual portions of the temporal lobe lie. In order to follow the course of the optic nerve as the relations on the skeleton indicate, we must direct the plane of section from the middle of the pupil, obliquely upwards.

The idea of this proceeding had already struck Sömmering, and in his monograph (' De Oculorum Hominis de sectione horizontali commentatio,' Göttingæ, 1818) a very useful plate is given, so also in Pirogoff's atlas (fasc. i, tab. iii, 4, 5). As follows from Sömmering's work, he employs the

* These terms are in frequent use in this work, and imply respectively a vertical section by a plane passing through the structures in an antero-posterior direction, a section produced by a plane also passing vertically, but at right angles to the former, whilst a transverse section implies one made at right angles to the axis of the trunk or extremities.—TR.

Tab. III.

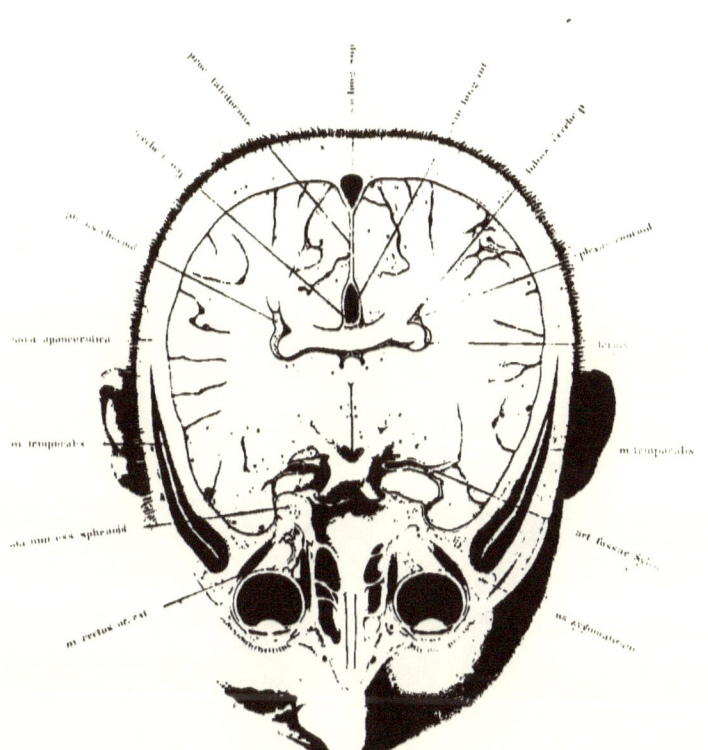

PLATE III   39

horizontal only as opposed to the longitudinal section, without thereby meaning that he adhered to a horizontal plane which was accurately mathematical. I considered it profitable to conduct the section in the same way as Sömmering has done, and to carry it up tolerably far back. I convinced myself, however, from a large series of sections, that we are not positively able to determine in what position of the optic nerve or tract the saw will fall from the front inwards. The individual differences of the base of the skull are so numerous that it is impossible to assign to them any definite data.

Only so much is certain, that the optic tract rises up from the chiasma to the corpora quadrigemina considerably more vertically than the optic nerve does to the optic foramen. I was therefore unable to expose the optic nerve thoroughly throughout its entire length, but was obliged to supplement the section by taking off a thin slice of the anterior lobe in order to completely expose the chiasma. Again, a thin layer of fat was removed from the orbit so as to show the entire breadth of the nerve, as the line of section had just missed its upper edge.

It must be further explained that, although the external form of the globe be established, the relations of the lens and iris must be rendered after further sections. The fine dust which even such a thin saw produces was very difficult to remove without causing a change in the relative position of the individual organs of the eye. I therefore froze fresh orbits, sawed through the bones, and then continued the section with a razor. In all cases the eye was thoroughly injected with Thiersch's carmine and glue preparation, in order to give the globe its original expansion. The injection was made from the ophthalmic artery, and in the entirely divided skull which forms Pl. III, the carotid artery and jugular vein were completely injected with different colours.

It will be noticed from the relations of the brain that the plane of section is obliquely upwards and backwards. In front, owing to the removal of the thin lamina of the anterior lobe, a small portion of the floor of the skull in the region of the crista galli is seen. Behind it are the optic chiasma with a small piece of the optic tract cut obliquely, and further back is a section of the gyrus fornicatus, the superior processes of which lose themselves in the white substance of the cerebrum, and show the small bundles of fibres, the beak-shaped processes which belong to the fornix below.

Externally are the choroid plexuses of the descending cornua. Beneath the white substance of the corpus callosum is a fissure bounded laterally by the optic thalamus and filled up with vessels, and in the middle of it is the pineal body. In this space also is the pia mater passing beneath the corpus callosum to the central portion of the cerebrum. In the middle are the lumina of two large vessels belonging to the great internal veins of the brain, the venæ magnæ Galeni. These, when followed down with the sound under the splenium to the great veins behind the corpus callosum, and the commencement of the straight sinus, are found to debouch by the two veins here shown. The falx cerebri unites the inferior longitudinal sinus which opens into the straight sinus, with the superior longitudinal which lies further back.

The veins of Galen curve at first upwards, and then bend downwards into the straight sinus. It would be worth while to examine whether during the movements of the brain in respiration any influence is exerted on the venous circulation at this spot, since a change here takes place between the compressible inner veins of the brain and the rigid walls of the sinus.

In front are the optic chiasma and the optic tract cut obliquely, which consequently are more vertical than the plane of the section.

Internally are the grey-edged lentiform nuclei, and behind them lie the sections of the optic thalami. Between them in the middle line is a portion of the third ventricle.

On either side of the chiasma are the internal carotid arteries in section. The ophthalmic arteries are not seen, as they enter the optic foramina beneath the optic nerves. The commencement of the artery of the corpus callosum on the one side and on the other is removed, the lumina seen being those of the artery of the fissure of Sylvius.

The orbits were so divided that the saw passed above the optic foramina, and consequently did not open them. The eyelids were tolerably firmly closed, so that only a small portion of the upper lid came into the section,

PLATE III 41

and the lower remained untouched. The globe was divided almost exactly in the middle. The saw cut the upper border of the entrance of the optic nerve of both eyes. In the orbits it was necessary to remove cautiously a thin layer with the knife in order to expose clearly the optic nerves. These show a slight curve, which on further examination is seen to be associated with a bending downwards, in the form of the letter S, which is their normal position while the eye is at rest. This fact renders it possible that considerable traction on the globe may exist without the optic nerve being stretched in any injurious manner in the various movements of the eyes. A tension of the optic nerves must occur unless they entered the optic foramina in this curvilinear manner. It may be left undecided, and for further investigation, whether by a stretching of the sheath of the optic nerve a lymph motion is brought about in its course. The researches of Schwalbe have established, as I may here point out, that the different tensions of Tenon's capsule during the action of the muscles of the eyeball may perhaps act in such a way. It is certain that this relative length of the optic nerve is necessary to ensure the position of the globe when the eye is at rest, and if the optic nerve were tense as it passed from the optic foramen to the globe it would be drawn out of this original position by the constant movements of the eyeball.

The optic nerves, which are represented rather too broad in the plate, are about ·16 of an inch in diameter. Here they are considerably narrower than when they are within the skull. In the latter situation they are from ·2 inch to ·24 inch broad. These diameters, however, appear at first sight to be equal on account of the slight length of the nerves. The orbit was not opened to its apex, and is, moreover, proportionally small even for a young person's skull. The length of the nerve from the optic foramen to its entrance into the globe is, according to Henle, about 1·2 inch, according to Sœmmering ('De Oculorum Sect. horiz.,' Göttingæ, 1818) 1·4 inch; in the present plate it is only 1·12 inch.

On the other hand, it shows a complete agreement with the statement of Henle that the centre of the point of entrance of the optic nerve is ·16 nearer the middle line from the posterior pole of the axis of the globe. Those portions of the globe which are shown in the preparation require no

6

further explanation. Their symmetrical form appears remarkable, as, according to the statement of Brüche, some considerable want of symmetry exists which is characterised by the fact that the equatorial plane, through the iris, lens, and ora serrata, converges to the nasal side. The globe, together with the cornea, exhibits much more nearly a circle in a section in the horizontal meridian, where the long diameter exceeds the transverse one almost imperceptibly.

It must, however, not be forgotten that preparations such as the present one cannot furnish in this respect any absolute standard. The numerous vessels of the choroid have, as is known, a considerable influence on the form of the globe. As I could not accurately measure the pressure of the injection, and as it was chiefly calculated to throw as much fluid as possible into the vessels of the eye, it is quite possible that it was considerably more powerful than it would be under natural conditions; great pressure in the vessels of the eye tends to approximate the form of the eyeball to that of a sphere.

It is to be further remarked that freezing does not seem to be the most suitable method of hardening the eye. Influences are thereby exerted which may bring about variations of volume in the watery contents of the globe; however, I know of no other method. It is here not so much a question of determining the forms and position of individual portions of the globe as it is of representing in section the relations of the eye to the orbit and to the other portions of the skull.

The dark fissure in front of the globe represents faithfully the expansion and folding-in of the conjunctival sac. Behind are the attachments of the external and internal recti, which are inserted into the globe beyond its axis of rotation—relations which are not represented correctly by Sœmmering. It is also evident that the internal rectus is inserted further forwards than the external.

At the inner angle, and lying on the lachrymal bone, is the section of the lachrymal sac, and on the outer angle between the muscle and the bone a small portion of the lachrymal gland.

The relation of Tenon's capsule cannot be represented on account of the small size of the plate; it was considered that a multiplicity of lines would

PLATE III   43

interfere with clearness.   Moreover, the relations of this membrane are not sufficiently made out.

It can be shown that this membrane is in relation with the tarsal membrane, and, with the mass of fat which lies behind it forms a sort of cup, into which the globe is pressed by the influence of the external air. Thus the eyeball moves somewhat like the head of the thigh bone in its socket, the fluid in the lymph spaces playing the part of the synovia. It is not yet shown how the capsule of Tenon is continued over the sheaths of the muscles where they pass back through it, still less so how the sheaths of cellular tissue stand in connection with it.   Such a demonstration would especially be of practical importance as regards the question of the extent of the effusion of blood in the orbit.

PLATE IV

THIS plate does not represent a directly transverse plane section, but an oblique one. In order to bring the relation of the ear well into view the section is taken obliquely backwards and upwards. It commenced close under the nose, and has involved in its course to the external meatus the inferior turbinated bone, the upper portion of the pharynx, the right Eustachian tube, the tympanum and the external meatus, and passing out through the pons Varolii, has divided the upper half of the cerebellum and the posterior lobe of the cerebrum above the external occipital protuberance. The preparation was made from the body of a young man, which presented nothing abnormal. As the section in the posterior half of the left side passes higher than in the right, it is nearer the roof of the left tympanum, whilst on the right it approaches its floor. I convinced myself, however, from many sections, that if there be any deviation of the saw from the proper direction it is not possible to make a thoroughly symmetrical preparation.

In this plate the objects of chief importance are the relations of the right organ of hearing, the section of which has been so fortunate, that not only the external meatus and the tympanum, but also the first part of the Eustachian tube as well, have been divided. I was never again able to obtain the parts together to such an extent, although I made more than twelve sections in the same direction. Owing to individual variations in the base of the skull in the direction of the Eustachian tube, it is impossible to give any exact definitions for making such a section.

As the chief points of interest lie in the upper half of the section, it will be represented instead of the lower half as in preceding plates; consequently one looks from below upwards into the skull, and the parts lying on the right side are really those of the left side and *vice versâ*. In

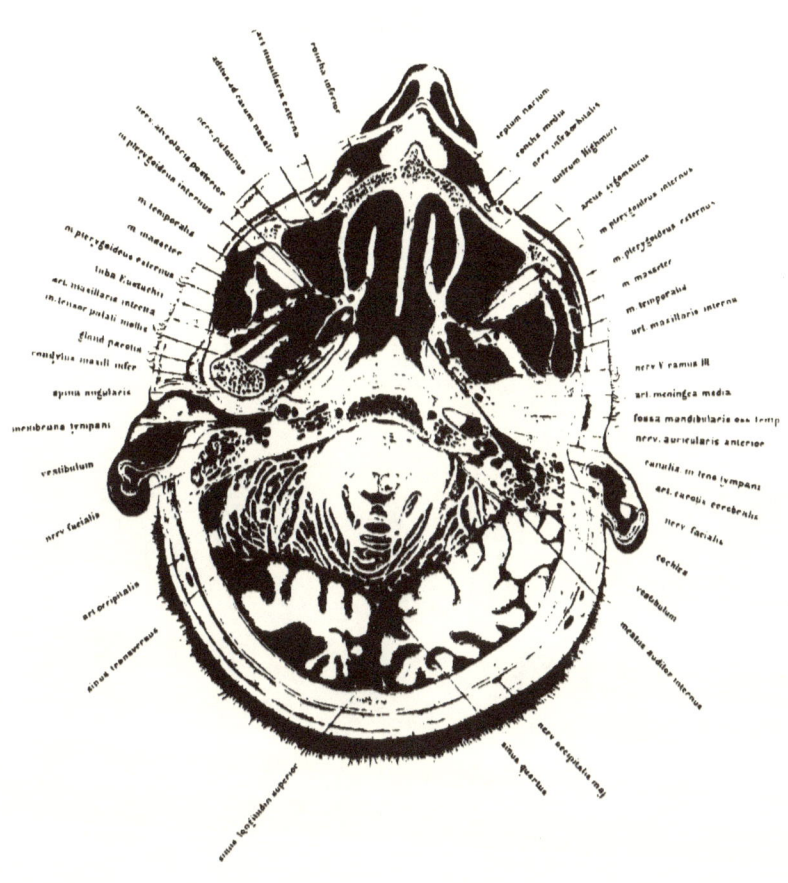

PLATE IV 45

describing the individual organs, then, of the left ear, the left nasal cavity, and so on, the right side of the plate must be consulted.

The upper half of the external meatus, the relations of the cartilaginous, integumentary, and bony parts of the right ear are seen. The connections of the cartilaginous portion of the Eustachian tube with the cartilage of the pinna, and the fissures of the external meatus appear as gaps between the cartilaginous rings.

By this disposition a large amount of passive motion is allowed at the entrance to the ear, which is noticeable equally in the movements of mastication and in the contractions of the pinna. By drawing back the ear the curve of the meatus is diminished and the examination of the membrana tympani rendered more readily accessible.

It is worth noticing that the curvature of the external meatus is not so abrupt as one might imagine from the examination of soft preparations or plates.

After examination of the living body as well as Pirogoff's plates (*a a O*, fasc. i, Tab. 6) which also were prepared from frozen bodies, and in spite of the various differences which in this respect the external meatus offers, I must admit that the parts seen in the section of a body not thoroughly hardened, vary materially with their original position. It is well known that this canal is curved from before backwards and from above downwards, and is thus somewhat serpentine in its course. Naturally, the relations in this section could not be represented with perfect clearness, although it will be seen from the stronger shading of the internal parts, that the upper wall of the canal rises backwards and somewhat upwards, and consequently that the semi-canal in the region of the membrana tympani is deeper than it is wide externally.

The membrana tympani has been divided in its lower half, hence the ossicles are not interfered with.

The direction and position of this membrane should be noticed, as it lies in a plane which makes a very acute angle with the horizontal, and also the navel-shaped retraction towards the tympanum, and the portion of the malleus which is in relation with it. As the ossicles of hearing, on account of their small size, are very difficult to represent accurately, a woodcut is introduced at the end of this chapter, in which the parts are enlarged three times. Deep down in the tympanum passing from before backwards is seen a bony protuberance which belongs to the semi-circular and Fallopian canals, and behind the stapes, in the section of the temporal bone, is shown the canal containing the facial nerve. This marked "cropping up" of this canal on the roof of the cavity of the tympanum is characteristic of the young individual. The young ear is especially suitable for the study of the organ. In the middle line from the stapes is the vestibule. The cochlea is not shown as it lies above the section.

Internally and in front of the cavity of the tympanum is the internal carotid artery, shown from its entrance to its first curvature, and subsequently divided transversely. In front of the carotid is the Eustachian canal, flat anteriorly, and passing deep down posteriorly. It runs consequently more vertically backwards than the plane of section passing from the nasal to the auricular aperture. The section passes through its pharyngeal opening, laying that portion of its canal free, but not its bony portion. Thus, only a small portion of the lateral tubal cartilage (Rüdinger's hook) is divided anteriorly, whilst a long strip of the median cartilage is exposed. Laterally the canal exhibits a mucous membrane rich in glands and cellular tissue. A portion of the tensor palati can be seen, and its origin can be traced backwards to the spine of the sphenoid. No portion of the levator palati is shown, as the section passed above its origin. It has been repeatedly proved that the tensor palati is at the same time a dilator of the Eustachian tube, and an elevator of the lateral cartilage, thus opening the canal.

The excellent representation of the Eustachian tube in Rüdinger's atlas ('Atlas des menschlichen Gehörorganes,' München, 1867) should be compared with this.

The projecting lip is clearly seen where the middle of the tube stands out at a point about ·6 of an inch from the posterior wall of the pharynx, and behind it is the fossa of Rosenmüller (*recessus infundibuliformis* of Tourtual).

The mucous membrane of the pharynx is rich in glands, and is continuous with that of the Eustachian tube and nasal cavities. It presents numerous blind crypts and depressions, which can be only hinted at in the

PLATE IV   47

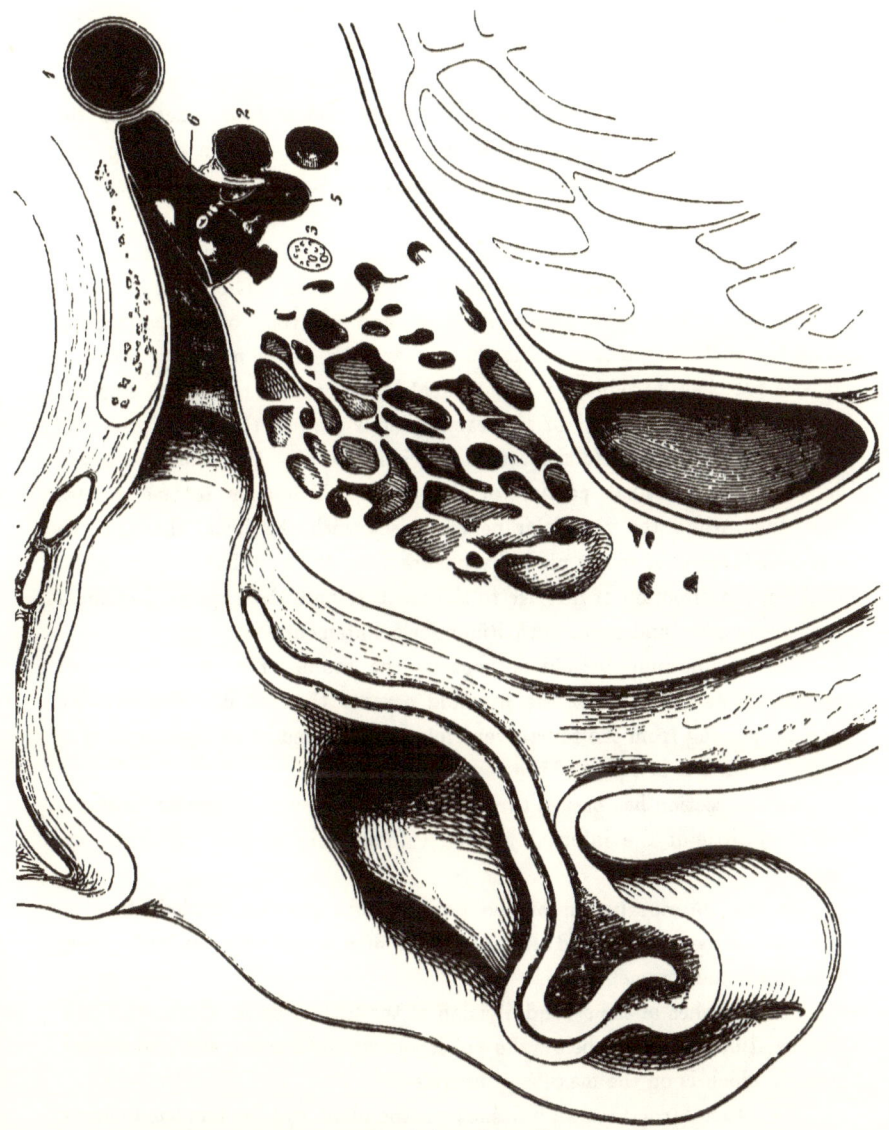

Section of right ear, enlarged three times. Seen from below.
1. Internal carotid artery.   2. Vestibule.   3. Facial nerve.   4. Corda tympani.
5. Stapedius.   6. Tensor tympani.

plate. The mucous membrane has fallen into the section at the point where it passes over to the roof of the pharynx, above the rectus capitis anticus.

From the relative position of the Eustachian tube to the pterygoid process and inferior turbinated bone, it is evident that œdema of the mucous membrane may easily close up its opening. Such a swelling may happen from cold, and nasal polypi are often the cause of difficulty of hearing.

As regards the left ear there is little to say, as the saw passed at a considerably higher level than on the right side. The cavity of the tympanum is laid open nearer its roof, and in front of its connection with the posterior portion of the Eustachian tube, at the middle of which a bristle has been introduced into the canal for the tensor tympani. Further forward is the upper half of the cartilaginous part of the canal. By the laying open of the left meatus auditorius internus, the auditory nerve is well shown. That portion of the nerve which goes to the cochlea is divided, while the vestibular nerve passes with the facial through the superior fovea (and in the plate disappears deep down). The section of the cochlea, the direction of its base to the meatus, and the exposed vestibule are clearly seen, and agree with Rüdinger's statements.

There is nothing to add as regards the brain. The pons Varolii in section shows the fibres of the pyramid passing through it. The anterior portion passing from the fourth ventricle to the aqueduct of Sylvius is met with; behind it is a part of the vermiform process.

As the section has passed through the skull above the jugular foramen, but very little of the internal jugular vein and eighth pair of nerves are seen.

On the anterior border of the pons Varolii are the divided fibres of the sixth nerve. The third division of the fifth is met with on both sides, just below the foramen ovale.

The branches of the second division of the fifth lying in the section are the palatine, which lie below the spheno-palatine foramen, and the dental branch which is on the maxillary tubercle.

The other structures and tissues on the plate will be alluded to individually.

Tab.V.

Fig I

Fig II.

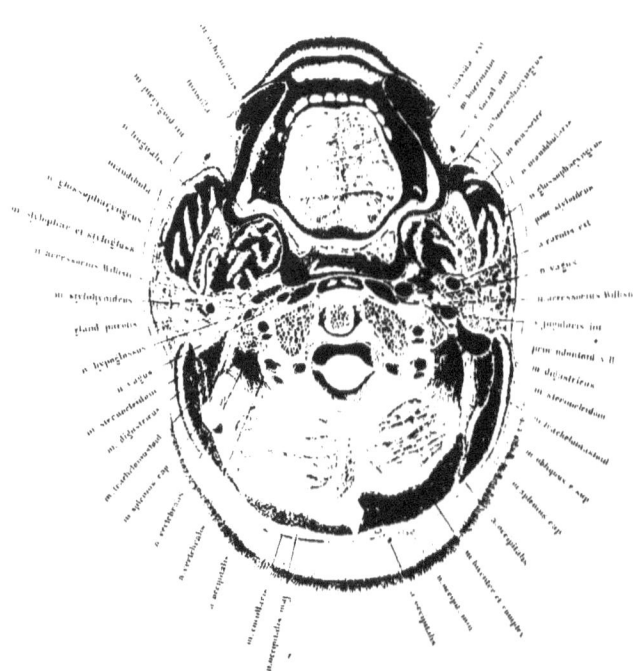

PLATE V

THIS plate and Nos. 6, 7, and 8, are made from sections of one and the same body. The region of the neck was cut in five series of planes of which the upper surface of each is represented and analysed, it is viewed from above downwards. The right side of the drawing is the right side of the preparation. Owing to this cutting into planes, the explanation of the individual outlines becomes considerably more difficult than had the sections been made on different bodies. By making very thin sections the arrangement of the muscles of the nape of the neck were very difficult of definition. On the other hand, this proceeding affords the great advantage that the under surface of each section fits exactly on the upper surface of the one next following it, and also that the separate organs, such as the thyroid body and larynx, which show pretty considerable individual differences with reference to size and position, can be analysed by transverse sections which mutually correspond. The body, which was of fine proportions and perfectly normal, was quite fresh, and was about twenty-five years of age. The muscular development was good. After the arteries had been injected, the trunk, from which the lower extremities were removed, was frozen in the usual manner, with the arms close to the side, and prepared as has been before described.

In consequence of the great muscular development the shoulders were very high, and therefore the neck appears to be comparatively short. It is not, then, to be wondered at that sections at the level of corresponding vertebræ in respect to the region of the shoulder, differ from those represented by Pirogoff (fasc. i, tab. ii), which were made from a person less thoroughly developed.

Fig. 1 corresponds nearly with Pirogoff (fasc. i, tab. lx, fig. 1), and

7

Henke (taf. lxx, fig. 2). The section passes through the mouth and runs somewhat above the level of the teeth, falling upon the hard palate and the lateral masses of the first cervical vertebra, and slices off a thin lamina of the cerebellum on the posterior edge of the foramen magnum. It is seen on comparing it with Plate I that the section passes obliquely backwards and upwards,. from the head being bent somewhat backward in the recumbent position of the body. This relation must be borne in mind in observations on the living body. In the normal upright position of the body a plane section through the level of the teeth would pass through the second cervical vertebra, and would not touch the skull at all.

After cleansing the preparation it appeared that a small portion of the dorsum of the tongue also was sliced off with the crowns of the upper row of teeth. The apex of the tongue remained just behind the teeth. Posteriorly the section had passed 1·5 in. from the foramen cæcum. The papillæ which are seen on the posterior portion of the section correspond to the middle of the tongue. In the middle line from before backwards is the septum linguæ, from which the fibres pass to both sides of the transverse muscles; in the posterior third the upper longitudinal fibres are seen. Behind the back of the tongue the uvula is retained in its entire length, as the section passed through the soft palate a quarter of an inch above its root, where the pillars of the fauces meet. The upper portion only of the tonsil is divided. In front of it, and behind the glands of the soft palate, lie some muscular fibres which pass transversely upwards, belonging to the upper border of the palato-glossus, which is embedded in the anterior pillar of the fauces. The azygos uvulæ is also seen. Behind the tonsil and in relation with it is the palato-pharyngeus, which forms the posterior pillar of the fauces. The accurate division of the muscles cannot be defined, nevertheless it appears as if the transversely divided bundle of muscular fibre behind the tonsil (especially the left), belongs to the levator palati. There is no portion of the tensor palati seen since the section passed below the hamular process. A portion of the superior constrictor of the pharynx is very well shown in connection with the lower jaw and the buccinator.

PLATE V   51

Within this muscular zone is the cavity of the pharynx. This space is often thought to be larger because it is observed on the living body in an oblique direction, through the posterior palatine arch ; in a vertical section made on soft preparations the interval between the uvula and the posterior wall of the pharynx is generally represented as far too large. In the operation of staphyloraphy one is often disagreeably surprised at the narrowness of the locality, and must resort to some one of the ingenious needles which have been invented on account of this want of room.

Behind the muscular tissue of the pharynx and the lax cellular tissue which in the plate is represented as a white line, lie the longi colli and recti capitis antici majores, and further outwards on the transverse process of the atlas is the tendinous origin of the rectus capitis lateralis. The position of the internal carotid artery is of the utmost importance in operations on the tonsil and pharynx. It is seen that this large arterial trunk lies in immediate relation with the muscular tissue of the pharynx; its pulsation can be easily felt from that cavity during life, and deep incisions in this region should not be made without the greatest caution.

The actual position of the artery to the tonsil, on the other hand, permits of greater freedom in extirpation of this gland, and numerous operations on it have shown that Hyrtl's apprehension (' Top. Anat.,' 1, 380) in this respect is much exaggerated. Nevertheless the proximity of the carotid must be especially borne in mind, even in the forcible dragging of the gland from its bed, but from the benign nature of most of the tumours of the tonsil there is no necessity for endeavouring to remove the gland completely, as the surgeon may be thoroughly satisfied if the chief mass of the growth be extirpated. As most of the instruments used for operations in this region only permit of a levelling and not of an extirpation of the tonsil, there is a sort of guarantee against wound of the carotid.* The position of the inferior dental and lingual nerves with regard to the lower jaw, is well

---

* By the use of a simple curved bistoury and vulsellum forceps the tonsil can be more readily and easily removed than by any other method, and the object of dragging the gland forcibly from its bed towards the mesial line, with a view of avoiding any chance of wounding the vessel, can be well recognised from examining the plate.—Tr.

shown. With regard to the latter nerve it is to be remarked that wounds of it in clumsy extraction of teeth from the slipping off of the instrument have often occurred. Its division in the mouth in neuralgia, as Roser has recommended, is thoroughly practicable, without cutting through the cheek. After extraction of the last upper molar the nerve may be divided with a tenotome on the ramus of the jaw, without the necessity of further dissection. The articulation between the axis and atlas is such that the saw has entered under the anterior arch of the atlas, dividing its odontoid process, and has met the occipital bone over the posterior arch. The powerful transverse ligament of the atlas passing obliquely behind the odontoid process, is separated from the bone by a bursa. Further back is seen the broad ligamentous mass of the lateral axoid ligament, which terminates on the body of the axis, and which partly passes over into the posterior common ligament. I have been unable to find in this preparation the synovial membrane described by Luschka as existing between these ligaments. Unfortunately, the two ligaments are not clearly enough defined from each other in the plate; the lateral portions of the ligamentum latum are represented too streaky. On the anterior aspect of the odontoid process lies the ligamentous mass filling up the space between the bodies of the axis and anterior arch of the atlas— the deep anterior axo-atloid ligament. A portion of this ligament is met with below the anterior arch of the atlas ; the anterior articular cavity lies above the section. It will be seen from the breadth of the ligamentous mass that the position of the odontoid process acts as a safeguard against powerful strains, and that the lateral masses of the atlas must have a great expanse in order to afford sufficient attachments for such strong ligaments. The mass of cellular tissue which closes up the space between the posterior arch of the atlas and the occipital bone is very lax, and the posterior atlanto-occipital ligament, which is cut at a very acute angle, takes up a considerable space. Immediately beneath, the posterior arch of the atlas can be felt. It is here that the vertebral artery makes its way, in order to perforate the dura mater further internally, and it thus reaches the medulla oblongata. The artery is divided three times, on account of its curves. The first section is in the vertebral canal, where the artery passes

PLATE V 53

vertically upwards, and the second where it bends backwards after the completion of the curve as an arc flattened transversely towards the middle line.

There is nothing particular to observe with regard to the section of the skull and the small lamina of cerebellum. As the skull was divided very superficially the prominences appear larger than they really are, and they thus acquire such singular forms. The muscles, vessels, and nerves of this region are readily recognised in the plate, and require no particular remark. The occipital artery of the right side is seen through a considerable portion of its length. It arises from the posterior aspect of the external carotid, and passing at first vertically upwards, crosses the internal jugular vein to reach the posterior belly of the digastric. From thence it runs horizontally backwards in the lateral region of the neck, being covered by the trachelo-mastoid and splenius. Having arrived at the middle edge of the splenius it pierces the upper origin of the trapezius (cucullaris), and then runs superficially on the skull. Very little of this vessel is to be seen on the left side. Between the splenius and the occipital bone a muscular branch appears passing deeply from its trunk.

The glosso-pharyngeal, vagus, spinal accessory, and hypoglossal nerves are shown in the plate.

The parotid gland is of especial surgical interest; it is enclosed in a dense fascial envelope which surrounds it on all sides, sending a multitude of septa into the substance of the gland, which account for the lobulated appearance which it presents on section. As the fascia lines the entire niche in which the parotid is imbedded, there is not only a demarcation between it and the internal jugular vein (which must especially be considered in the extirpation of tumours of the gland), but it is a protection to the vagus, spinal accessory, and hypoglossal nerves, which are in close proximity. The portion of the fascia which is most strongly developed is that which covers in the outer aspect of the gland. This, on account of its connection with the fascia of the masseter, is called fascia masseterico-parotidea.

In consequence of this arrangement the swelling of the gland from inflammation is limited externally, thus the tumour presses internally against the nerves and vessels.

The parotid being pierced by the terminal branches of the external carotid artery and the posterior facial vein, its extirpation without wounding these vessels is impracticable. But if the carotid, as shown on the right side of this preparation, lies so peripherally that it can be dug out of the mass of gland tissue, an operation would be less uncretain in its result.

On account of the numerous anastomoses of the arteries in the skull it is of little use to attempt to arrest hæmorrhage from the external carotid, but in the event of complete extirpation of the gland it would be necessary to direct one's attention especially to the preservation of the internal jugular vein.

Figure 2 represents the upper surface of a lamina 1·5 inch thick, which corresponds with the under surface of Plate IV. The section which the plate represents has passed through the thyroid notch and has fallen close on the upper border of the fifth cervical vertebra.

As the section passed immediately below the chin and lower jaw, the neck would be divided in its so-called cylindrical portion. It is seen, however, that on account of the muscular development at this level the natural form of the neck is not an exact cylinder, consequently its section is not a circle, but is a pentagon.

The lateral portion of the trapezius (cucullaris) commences immediately below the section, consequently the plane of section is enlarged, corresponding with the anterior curvature of the cervical spine. The section of the vertebra lies much further removed from the side of the neck than one would expect. In the accompanying plate the body of the vertebra lies in the anterior half of the figure. The point met with in the section of the vertebra is where the arch springs from the body, hence the lumen of the spinal canal is seen. On the left side is seen the articular process of the sixth cervical vertebra, and from this point we can follow the course of the sixth cervical nerve behind and to the outer side of the vertebral artery. The divided nerve seen lying in the bifurcation of the transverse process is the fifth cervical.

The larynx is so divided that the vocal cords with the ventricle of Morgagni between them is clearly shown. The mucous membrane which lies behind the divided arytenoid cartilages is here singularly rich in glands.

PLATE V   55

In the section also are seen the arytenoid glands, many of which are embedded on the inner side of the aryteno-epiglottidean fold. The thyro-arytenoideus and the arytenoideus are divided through their upper extremities. From the arrangement of the muscles, the resemblance to a sphincter can be clearly recognised. Behind this layer of muscles lies the large mass of glands of the pharynx—the middle arytenoid gland of Luschka.

The common carotid artery is shown exactly in the position which would be most suitable for its ligature; its relations should be carefully noticed. It is at the spot where the deviation of the omo-hyoid and sterno-cleido-mastoid allows of a ready means of access.

However incompletely the relations of the fasciæ in such a plate are rendered, one can see clearly that the vessel must be sought on the anterior border of the sterno-cleido-mastoid, and that after the division of the hinder portion of its sheath the space containing the great vessels and nerves is immediately entered.

In front of the artery is the descendens noni, and somewhat behind it is the internal jugular vein   between the vein and artery is the vagus, and behind the artery is the sympathetic. Inside the sheath of the vessels a layer of cellular tissue isolates the artery from the vein and the nerve. The important part of the operation consists in opening the sheath of the artery, which lies immediately in front of the scalenus anticus muscle. If this be properly done, one not only avoids the danger of wounding the nerve, but the vein also will be kept at a distance. After opening the sheath the vein would expand enormously and cover up the whole field of operation.*

* In ligature of the carotid pressure should be made on the internal jugular vein by the fingers of an assistant both above and below, in order to prevent this expansion of the vessel from interfering with the operator's movements.—Tr.

PLATE VI

This plate is taken from a section of the same body as the last, and has been prepared in the usual manner.

The section passed through the larynx, and should properly have kept to the plane of the lower vocal cords, but it passed above them in a horizontal direction, and fell on the lower half of the sixth cervical vertebra.

The body has a peculiarly well-arched thorax, and owing to the great muscular development, the shoulders are high up, and although there are the normal number of vertebræ the neck appears short, corresponding in the most marked degree with the male type of neck formation. Here again the section does not show a circular contour, but rather a prismatic one. It is easily seen that this is owing, to a great extent, to the powerful muscular development of the sterno-cleido-mastoids and the trapezii.

As the section has not passed through the head of the humerus, but through the acromio-clavicular articulation, it did not traverse the shoulders at their greatest breadth, but at the junction of the regions of the neck and shoulder. Therefore the lateral portions of the plate represent only the upper portion of the roundness of the shoulder, the supplementary parts of which will be shown in following plates.

The slight irregularity noticed in the edges is owing to loss of substance after the use of the saw.

In the female, or slightly developed male subject, the lamina, which in this case was about ·4 in. thick, would have taken a totally different form, as the position of the shoulder would be lower in the so-called cylindrical

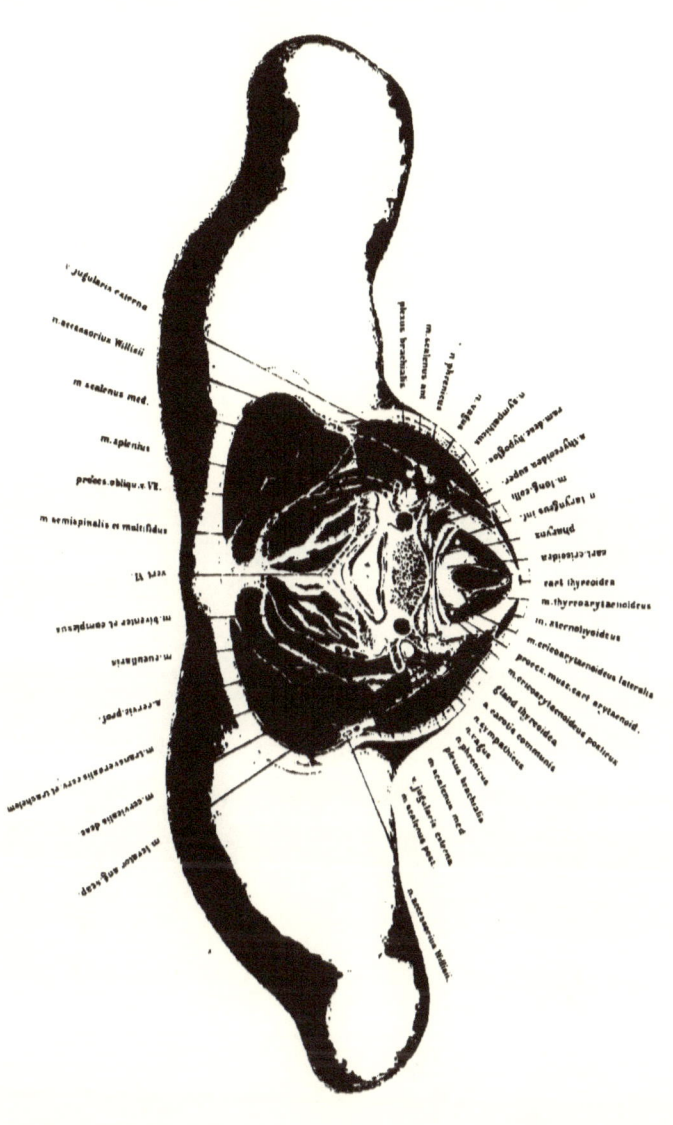

jugularis externa

n. accessorius Willisii

m scalenus med.

m.splenius

proces.obliq.v.VI.

m semispinalis et multifidus

verv. VI

m.bicentice et complexus

m.rectus maj.

m.rectus.post.

m.intertransvers.cerv.et trachelom

m.cervicalis desc.

m. levator ang. scap.

n phrenicus

m.scalenus ant.

n.vagus

v.jugularis int.

m.sympathicus

m.thyreohyoid.super.

m.long.colli

n laryngeus inf.

pharynx

cart.cricoidea

cart thyreoidea

m.thyreoarytaenoideus

m. sternohyoideus

m.cricoarytaenoideus lateralis

m.crico.must.cart arytarnoid.

proces.must.cart arytarnoid.

gland.thyreoidea

a. carotis communis

sympathicus

n.phrenicus

plexus brachialis

a scalenus med

v.jugularis externa

v.subcl.post.

a.accessorius Willisii

PLATE VI   57

portion of the neck, and consequently exhibit no lateral expansion in
the region of the junction of the shoulder and neck, the upper surface
of such a section, however, would offer another shape, and approximate
more to the circular. Pirogoff's plate (fasc. i, tab. x, fig. 5) should
be examined in order to prove that it is the feebly-developed muscular
neck which takes the circular form. Pirogoff, moreover, says that
his section was taken from an *emaciated* body; however, I maintained
from recent sections on a man of fifty years of age (such as is represented in Tab. ix of the first volume), that a section at the level of the
sixth cervical vertebra is tolerably round. The present case must then
be regarded as typical of the neck of a young powerful male, and deviations
towards the circular form on the living body are to be referred to want
of muscular development.

Sections on unhardened bodies naturally give no fixed forms corresponding with their original relations. The parts yield so much
on bodies which have been frozen and subsequently thawed that the
neck gradually acquires a circular shape. This may very likely be the
reason that the plates of Beraud and Nuhn, which represent very
similar regions of the neck, differ so essentially from mine as regards
external form. (Beraud's plate is in his 'Atlas d'Anatomie Chirur-
gicale,' Paris, 1862, pl. xxxvii. Nuhn's is represented by Henle,
'Muskellehre,' p. 131, and by Henke, 'Abl. der Topographischen Anatomie,'
taf. lxix.)

As to individual portions of the present plate to be studied, the first of
all is the larynx, which is divided close below the vocal cords; anteriorly is
the arc, formed by the section of the thyroid cartilage, and close behind it
the section of the cricoid. Of the arytenoid cartilages only the muscular
processes are met with, and nothing is seen of the vocal processes, as they
lie higher. The space between the thyroid and cricoid cartilages is filled
up with the thyro-artenoideus and crico-arytenoideus lateralis. On the
other side are some fasciculi of the thyro-epiglottideus. Behind this
and on the anterior surface of the crico-arytenoidei postici lie the inferior
laryngeal nerve and artery.

From the form of the transversely divided trachea it will be observed

8

that the section does not pass far below the rima glottidis, and that the
surface of the cricoid cartilage is divided obliquely forwards and downwards. The space expands still wider further downwards, and changes its
laterally compressed form for that of a cylinder, as far as to the point where
the cricoid cartilage encloses it completely. Finally, in the trachea it
becomes in section a segment of a circle.

As the present plate offers no points of great interest as regards the
relations of the larynx, I have made on a preparation hardened in alcohol,
a section exactly in the plane of the vocal cords and introduced it in the

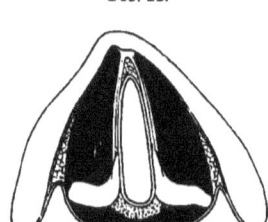

FIG. 11.

accompanying woodcut. It will be seen that
the processus vocales are continuous immediately with the elastic fibres of the vocal
cords. At the line of section, which is not
sharply defined, some reticulated cartilage
exists. In front the vocal cords pass into a
roll of connective tissue to which the thyro-
arytenoid muscles are attached. The mucous
membrane on the vocal cords is destitute of
ciliated epithelium, and is stretching tightly over and is firmly attached
to them. Beneath the mucous membrane the glands in this plane lie
in the angle between the anterior extremities of the vocal cords and
between the arytenoid cartilages posteriorly. On either side of the
vocal cords are seen the two cut surfaces of the thyro-arytenoidei, of
which the median is shown as internal and the lateral as external. Still
more externally are the cut fibres of a muscle which passes partly to the
thyroid cartilage and partly to the epiglottis, the thyro-aryteno-epiglottideus
(Henle). Behind the section of the arytenoid cartilage the arytenoideus
is seen in section passing across from one cartilage to the other.
Referring again to the large plate, we see behind the cricoid cartilage and
behind the section of the crico-arytenoideus posticus, the transverse chink
of the pharynx. The section shows it empty, therefore its anterior and
posterior walls are in contact ; behind it is the middle portion of the
inferior constrictor of the pharynx. As the pharynx lies immediately
upon the vertebræ, and the longus colli and recti capitis postici majores,

PLATE VI 59

the space required by the morsel of food in passing downwards is provided for by the dragging forward of the anterior wall of the pharynx and the advancement of the larynx. The larynx is, moreover, lifted in swallowing. The result of this twofold change in position is a movement of the larynx towards the chin, which can be easily observed during the act of deglutition. The lax cellular tissue which lies between the pharynx and the vertebræ appears in the section as a narrow border, and by its extraordinary looseness it permits of the movements of the pharynx upon the vertebræ. But it is of such a nature that hæmorrhage into it would cause great distension. This condition is, moreover, favorable to the infiltration of pus.

Behind the pharynx lies the section of the sixth cervical vertebra, which has been divided in its lower half. As the section fell to the right side, and exactly at the springing of its arch, a clear view is furnished of the lumen of the spinal canal, which has the form of an equilateral triangle, and is so spacious that in the most extensive movements of the cervical vertebræ the spinal cord has free room, and is thoroughly protected from strain.

The relation of the vertebra to the surrounding soft parts is worthy of notice, inasmuch as it appears to be pushed remarkably forwards. If half the diameter, for instance, be taken from before backwards, the body of the vertebra would lie completely in the anterior half of the section. By comparing the measurements with those of the section shown in Plate I, and also in the other figures, it is seen that this position of the vertebra is correct. This appearance is owing to the cervical curvature of the spinal column. The distance of the medulla from the surface of the neck on the living body is usually represented as far too slight. Very similar relations will be found in Pirogoff (fasc. i, tab. iii, fig. 2; tab. ii, fig. 1; fasc. i, tab. x, fig. 66).

As the body of the vertebra is cut through near its lower border, its connection with the transverse process is clear. The vertebral artery full of injection, with its satellite vein, is seen in the bony canal on either side. On the left side, the section has fallen rather deeper, so that the canal in the transverse processes is closed in posteriorly merely by ligamentous tissue; it involves also the superior articular process of the seventh cervical vertebra and its joint cavity. Since the body of the sixth cervical vertebra with its transverse process is divided, a proper opportunity is afforded of examining the so-called tubercle of Chassaignac and its relation to the common carotid artery. Among surgeons this process is known as Chassaignac's tubercle, and is considered, according to the statements of authors, to be a most valuable landmark in seeking the vessel, in cases where ligature is rendered difficult on account of swelling of the tissues or the presence of a tumour.

It is clearly seen that the anterior of the tubercles of the bifurcated transverse process, which proceeds from the side of the body of the vertebra, and encloses the sixth cervical nerve, is a direct guide to the common carotid artery, which lies immediately upon it. Further, with regard to this tubercle, it has a morphological importance as a rudimentary rib, and is correctly called the eminentia costaria, jutting out more markedly from the sixth vertebra than from any of the others. It can be readily felt in the living body if gentle pressure be made on the side of the body of the vertebra upwards towards the level of the larynx.

Although advantage may be taken of the presence of this tubercle in looking for the vessel, for the sake of demonstration, and of making beginners acquainted with its locality, still it is not necessary for surgeons of experience to avail themselves of such a means of assistance, even in complicated cases. If the vessel has to be ligatured exactly at this spot, it is better to make the usual dissection over the course of the artery, dividing layer by layer. In this way there is less danger of wounding important parts, whilst the course to the vessel is sure.

The position of the vessels is defined by muscles and fasciæ, but these can be easily pushed away from their relations with the bony points. When, however, the vessels lie in bony canals, and are enclosed as fixedly and unalterably as the vertebral, for instance, then undoubtedly the determination of their position is facilitated. But, on the other hand, the means of reaching them may be rendered proportionately difficult. As, however, the carotid can be easily drawn away from its relation to Chassaignac's tubercle,

PLATE VI   61

this prominence as a means of assistance is not directly suitable in all cases, as is already proved by examination of the normal thyroid body (see figure), the upper lobe of which lies between the artery and the thyroid cartilage. Swellings of this gland must draw the artery away from the bony prominence, but they do not permit of its being released from the strong fibrous sheath, which is formed by the investment of the sterno-cleido-mastoid and scalene muscles, and of the gland itself.

From a section which I made at a similar level in the neck on a well-frozen body affected with goitre, the carotid was half an inch external to the tubercle in question, but the relations of the muscle and fasciæ were unaltered. On a closer examination of the plate the relation of the fasciæ to the artery will be seen. It is true that such representations are insufficient; and in order to make clear the relations of all the fasciæ one is compelled to represent them as white lines. I have therefore been able to mark out satisfactorily the coalescence of the several laminæ. Moreover, actual fasciæ cannot be properly distinguished from layers of cellular tissue. For the more accurate relations of this part I refer to the works of Dittl, Pirogoff, and Henle. I may add that the contours of the muscles which chiefly determine the arrangement of fasciæ are sufficiently accurately represented in the preparation, and in this respect furnish trustworthy points of reference.

Externally, and somewhat behind the artery, is the internal jugular vein, and between these vessels is the vagus, which in ligature of the artery must be carefully protected from injury. It is most safely avoided, if after division of the fibrous sheath a fine director be passed through the cellular tissue immediately over the artery, and then the edges of the fascia pulled aside with two pairs of forceps, before passing the ligature needle. By this means the ligature can be as readily applied, either from without inwards or from within outwards. Behind, and nearer the artery, is the sympathetic nerve, which may be avoided if merely the old rule be followed with respect to the vagus, of introducing the needle from without inwards. Behind the vagus, and on the anterior scalene muscle, lies the phrenic nerve.

Behind the jugular vein, between the sterno-cleido-mastoid and the

middle scalene muscles, are the supra-clavicular twigs from the fourth
cervical nerve.

Between the anterior and middle scalene muscles are the sections
of the fifth and sixth cervical nerves, which are figured collectively on the
plate as brachial plexus, so as not to disturb the detail of the clearness
of the drawings. The seventh cervical nerve comes off from the spinal
cord in the vertebral canal, and takes a direction outwards and backwards
behind the vertebral artery.

The above-mentioned figures of Nuhn (' Chirurg. Anat. Tafeln.,' taf.
iv, fig. 2) and Beraud ('Atlas d'Anat. Chirurg.,' Plate XXXVII, fig. 2)
should be compared, as the question to be proved is whether in these plates
of sections of the neck the natural relations are represented, since they show
not round but polygonal contours. There is one word to be added here on
the relations of this section with respect to the vertebra, in order that no
misconception may arise :—Nuhn's section of the larynx is taken almost
at the same level as mine, whilst in Beraud's nothing of the trachea below
the cricoid cartilage is seen. Both authors make the corresponding
vertebra the fourth cervical, whereas in mine the sixth is shown. One
might easily conjecture, therefore, that I have represented a wrong
vertebra—an error which may be easily committed if one has been already
making many sections of the neck. I, however, expressly state that
I went to work most accurately in the definition of the vertebræ, and
believe that I have made no mistake in the accompanying plate.

By comparing the vertical sections on Pl. I and II, as Pirogoff gives
it, the fourth cervical vertebra is on the level of the epiglottis, and the
seventh has the flat surface of the cricoid cartilage in front of it, which
also in this particular agrees with my plate. It cannot be disputed that
other variations in this respect happen to the extent of the level of a vertebra.
These variations are in all probability occasioned by the different degree of
curvature of the cervical spine. Nevertheless, I do not think that this
change in position can be extended to two vertebræ, and I maintain that
Beraud's statement that the fourth cervical vertebra lies deeper than the
cricoid cartilage is not correct. There is a vertical section in Beraud's
atlas (Pl. XXVIII, fig. 2) which bears out my statement. Perhaps,

PLATE VI   63

therefore, the parts were pushed out of their places in making a section of a soft preparation.

Pirogoff's transverse sections of the regions of the neck (fasc. i, tab. x) coincide with my account.   The cricoid cartilage here lies in front of the sixth cervical vertebra.

PLATE VII

THIS plate and those which immediately precede and succeed it are taken from one and the same body. Here, as is evident, the superior surface is represented.

The section, commencing immediately below the larynx, passes through the under edge of the cricoid cartilage, and involves the lateral lobes of the thyroid body, the under surface of the seventh cervical vertebra, and a portion of the intervertebral fibro-cartilage. It terminates at the level of the articulation between the acromion and clavicle. As can be verified by measurement, the lateral halves are of equal length, and there is considerable symmetry in the arrangement of the individual portions, so that the track of the saw is exactly horizontal, and yet there are many differences on the two sides of the section. On the right side the part between the clavicle and acromion is opened, and a portion of the muscular mass of the serratus magnus crops up from the scapula; the head of the first rib is also plain. On the left side the section passes beneath the articulation between the clavicle and acromion, and neither the rib nor the angle of scapula can be seen. It will be observed, then, that in normal and faultlessly formed bodies deviations from lateral symmetry may occur, a fact which does not permit representations of sections of one half of the body only being made.

As the section passed through the point of junction of the cervical with the dorsal vertebræ, it represents the area between the back and the neck. In front of the vertebra the section keeps completely within the region of the neck which descends lower down in front than it does behind. The clavicle can be observed in considerable length through the integument.

This relation must be borne in mind in examining deep-seated gun-

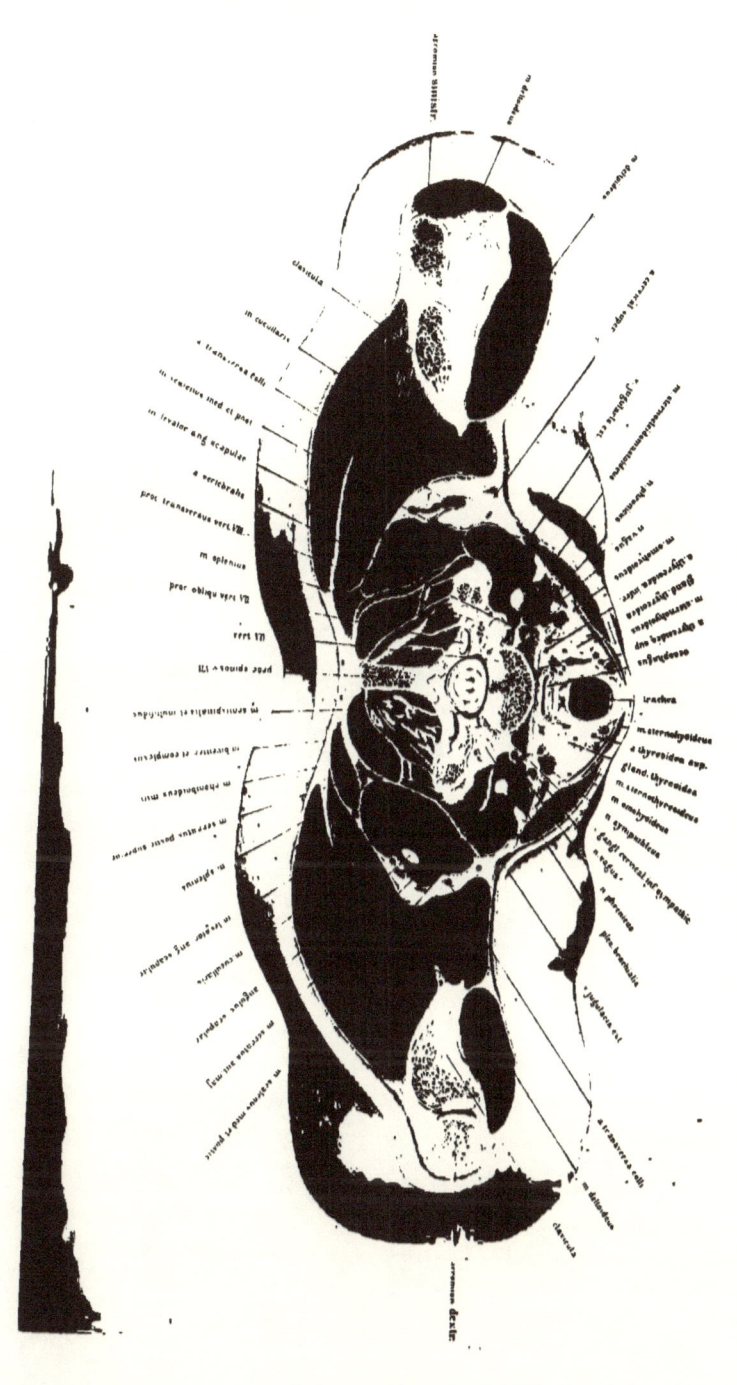

PLATE VII 65

shot or punctured wounds in this region. Students and beginners especially are liable to look for the highest of the dorsal vertebræ much
deeper in the neck than at the lower border of the larynx. Of the bony
portion of the vertebral column here seen we have the under surface of the
body of the seventh cervical vertebra, and its long spinous process; this
can be readily felt through the integument, and is useful as a commencing
point for counting the dorsal vertebræ. To this vertebra belong the
sections of the divided articular process. In front lie the articular and
transverse processes of the first dorsal vertebra, and on the right side juts
out the head of the first rib.

In front of the spinal column on either side of the median line is the
longus colli, and close beside it on the transverse process is the scalenus
anticus. The latter muscle is separated from the scalenus medius by the
transverse section of the brachial plexus, which is formed by the anterior
branches of the last cervical and first dorsal nerves; the posterior branches
are not clearly seen in the preparation. On the anterior surface of the
anterior scalenus is shown the phrenic nerve.

Between the longus colli and scalenus anticus lies the vertebral artery
with its vein, which have been divided on their way to the vertebral
canal; and immediately in front of the vein on the right side is the inferior
cervical ganglion of the sympathetic. On the left side the sympathetic lies
between the scalenus and the carotid, and a branch of the inferior thyroid
artery is to be seen on the inner side.

On the front of the spine is the trachea, which passes obliquely downwards and backwards, and its section shows the rest of the cricoid cartilage. Posteriorly the inferior constrictor of the pharynx indicates the
position of the commencement of the œsophagus, and as this canal is
empty, its anterior and posterior walls are closely approximated. Plate
VIII, which is taken at the level of the first dorsal vertebra, shows that
the gullet deviates considerably towards the left side.

Between the trachea and œsophagus on either side is seen the recurrent
laryngeal nerve, and more externally are the lateral lobes of the thyroid
body. Just at this point all the four arteries of the gland are visible, and
one can easily appreciate the difficulty of applying a ligature to them. The

superior thyroid artery has already entered the anterior portion of the
gland, and the inferior thyroid is seen external to it. On the right side,
between the carotid and the deep muscles of the neck, are two large divided
vessels which belong to the inferior thyroid artery, which springs from the
subclavian, passes for a while upwards in order to curve behind the carotid,
and then again downwards so as to reach the thyroid body. The vessel is
divided just below the loop, so that both its ascending and descending
portions are seen. On the left side the descending portion of the vessel
has already given off branches.

It has been before mentioned that in such sections as these there is
very great difficulty in representing fasciæ, consequently all the finer
laminæ lying between the different vessels have been omitted. The
space, however, in which the carotid, internal jugular vein, and vagus
nerve are enclosed is filled up with cellular tissue. The limits of this
space, as a whole, are accurately shown, and it can be understood that the
sheath of the vessel is formed anteriorly by the middle portion of the fascia
of the neck and the omo-hyoid, internally by the envelope of the thyroid,
posteriorly by the lamina covering the deep muscles, and externally by the
sheath of the sterno-cleido-mastoid. Inside the sheath, externally and
somewhat posteriorly, lies the vein, and in front, between the artery and
the vein, is the vagus nerve. The descendens noni lies on the front of the
sheath. At this point the omo-hyoid begins to cross the great vessels and
to become tendinous, as is well seen on the left side.

It is a recognised fact that the operation of tracheotomy should be
performed by preference above the thyroid body; between it and the cricoid
cartilage. The plate shows how near the surface the trachea lies in
this region, and how easy an operation on this body would be on account
of the normal condition of the thyroid body and the slight development of
its middle portion. As shown on the plate, one might be misled by the
relations, and look upon it as advantageous to perform the operation by a
single stroke of the knife; but the practitioner is warned against such a
proceeding; he must divide the tissues cautiously layer by layer, inasmuch
as the middle portion of the thyroid body may widely displace the
field of operation, whilst hæmorrhage from it is very difficult to arrest.

PLATE VII 67

The muscular masses which compose the posterior half of the central part of the plate may be thoroughly analysed, and they are sufficiently distinguishable from the references. It should be here observed that on the left side the section has passed above the serratus magnus, whereas on the right side its upper edge only is involved, and is in such close relation with the levator anguli scapulæ that no clear line of demarcation can be shown. On the left side, in the separation between the levator anguli scapulæ and the trapezius (cucullaris), and over the upper border of the serratus magnus, lies the transversalis colli artery, which is divided through its curve. Its course is clearly shown from the outer side of the scalenus medius, and its relation to the superficial cervical artery; up to this point it applies itself posteriorly to the trapezius and levator anguli scapulæ, in order to terminate in the region of the angle of the scapula; on the right side this vessel is seen merely in section. The trapezius forms the largest surface of the section and is divided just at the point of its fan-shaped expansion. The posterior fibres run more transversely to the acromion and acromial end of the clavicle, and are therefore cut parallel to their course; the mass of fibres lying on the anterior border are more perpendicular to the middle portion of the clavicle, and are divided almost transversely. The spinal accessory nerve is shown in this muscle.

PLATE VIII

Tᴴɪs plate represents the upper surface of the last section made from an uninjected body, which has also afforded material for the previous plates ; it is therefore unnecessary to mention anything further with regard to this body, as the essentials will be found with the explanation of Plate V.

The section is so made that it passes directly through both subclavian arteries at the level of the arches which they describe over the cupola of the lung, and very fortunately the trunk of the left subclavian artery remains intact ; whilst that of the right side is divided together with that portion of the lung which lies immediately beneath it. It extends moreover to the level of the isthmus of the thyroid body, to the lower edge of the first dorsal vertebra, and to the coracoid process of the humerus above the tuberosities. One consequence of the high position of the shoulders of this individual is, that the lateral portions of the shoulder-joint are seen in this section, whilst in the case of less powerfully developed bodies they are not met with till the level of the sterno-clavicular articulation is reached.

With regard to the relations of the spinal column, we first notice a small portion of the body of the under surface of the first dorsal vertebra, behind it the connexion between it and the second, which in consequence of the curvature of the spinal column has been sawn through obliquely. Small portions of the transverse processes of the second dorsal vertebra appear behind the intervertebral substance. The second ribs are seen attached to these processes and also to the bodies of the vertebræ. In the mass of muscle in front of them lie the first ribs in section. Nothing is seen of the sternum and the sternal end of the clavicle, since these

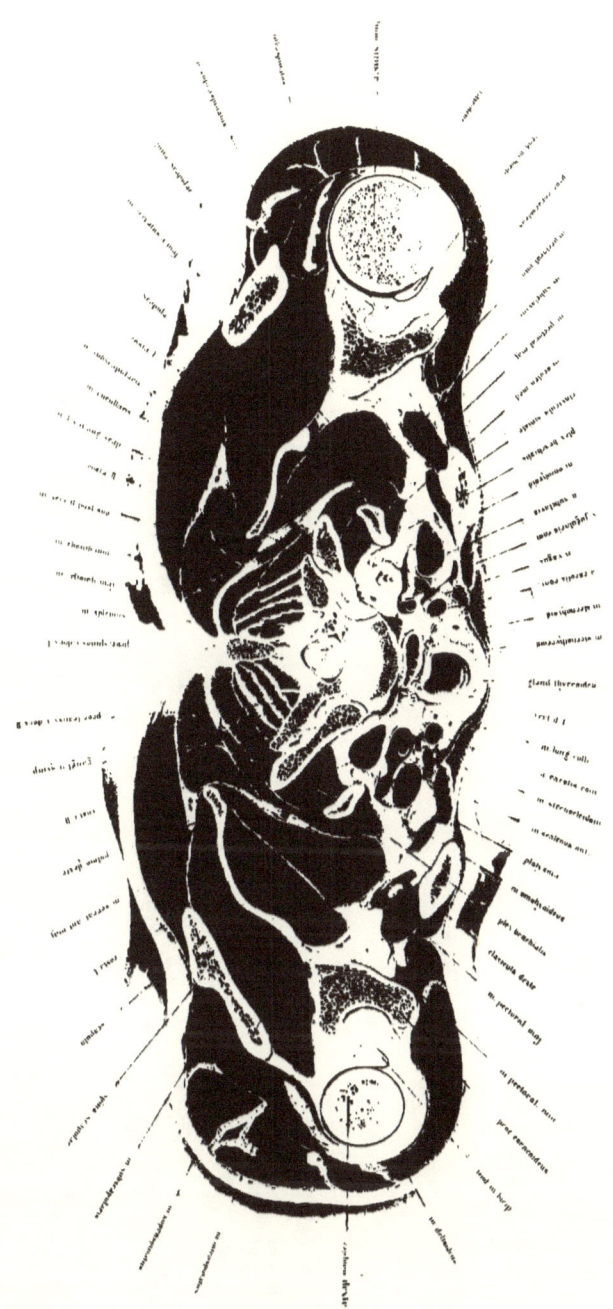

Tab. VIII.

PLATE VIII 69

parts lie considerably deeper, as is shown by an examination of the thyroid body. The section of the clavicles passes through their middle, and the subclavian muscles are readily seen. The upper portion of the thorax is opened by the section, which also implicates the region of the neck in front. Hence it is impossible to determine by means of a horizontal plane where the region of the neck terminates and where that of the thorax commences, but the boundary must be carried obliquely backwards, and even then the neck may be said to lie not only above the thorax but partially in front of it. Consequently it cannot then be wondered at that wounds penetrating the neck horizontally above the clavicle frequently involve the lung, and this fact must he kept in view in the examination and diagnosis of the course of stabs or gunshot wounds of the lower region of the neck. As the left lung is clearly seen through its exposed and uninjured pleura, whilst the right lung and the subclavian artery are divided, one might perhaps imagine that the saw had been depressed on the right side; this, however, was not the case. Although the horizontal plane was adhered to as accurately as possible, still the head of the right humerus has been divided at a considerably higher level than the left. We might suppose, therefore, that in this body the right lung attains a higher level than the left. This difference is clearly seen in the plate, and as it occurs in the case of a young and perfectly formed normal subject, it is obvious that this disposition is of importance in percussion of the apex of the lung. One would naturally expect in a young muscular individual a fuller percussion note above the clavicle on the right side than on the left, and if the reverse condition should present itself the existence of some abnormality may be expected.

On both sides of the muscular masses of the longus colli, between it and the lung, is the second cervical ganglion of the sympathetic; in front of and above the cupola of the lung is the subclavian artery, and laterally appears the obliquely-divided surface of the brachial plexus. As the artery does not exceed the highest level of the apex of the lung, but lies more on the anterior slope of the pleura, the brachial plexus forms a sort of niche with the spinal column to receive the absolute apex of the lung. On the left side especially this arrangement is well seen.

The left subclavian artery is intact, but sections of two of its branches are represented. The inner of these is the vertebral, the outer the thyroid axis. In front the superficial cervical artery winds round the anterior scalene muscle and the phrenic nerve, and mounts up obliquely above the brachial plexus in order to gain the nape of the neck. It has been divided at the commencement of its course, and immediately below; the posterior belly of the omo-hyoid overlaps it, a small portion of the muscle having been cut off, but almost the whole of it is shown in the section immediately preceding. On the hinder border of the subclavian are the openings of two small arteries which are not very clearly defined. The transverse cervical artery, the extremity of which is seen in the preceding plate, sprang, in common with the inferior thyroid, from the large trunk in the mass of the scalenus anticus, and the continuation of its trunk (the posterior scapular) is seen to be covered by the rhomboid muscle.

The supra-scapular artery lies behind the subclavian, and is again seen near the coracoid process, behind the conoid and trapezoid ligaments, whence it passes towards the supra-scapular notch. It runs over the transverse ligament of the scapula to the supra-spinous fossa, whilst the accompanying nerve passes below the ligament.

The section has removed a strip of the upper surface of the right subclavian artery, and at the inner end there is a bulging out of the wall of the artery corresponding with the origin of the thyroid axis, and indicating the point of origin of the superficial cervical artery. Further outwards, between the subclavius and serratus magnus are the supra-scapular nerve and artery to which we have already alluded.

On comparing the subclavian arteries of the two sides it is evident that on account of the higher level of the right over the cupola of the lung, that the first portions of both have very different directions. These differences are dependent on the variation of origin of the two vessels. The ascending portion of the left subclavian (from the aorta) lies further backwards, and is in relation with a considerable portion of the pleura, whilst the right passes in the opposed direction of the blood-stream, forwards, to unite with the common carotid to form the

PLATE VIII   71

innominata. The portion of each artery here shown belongs to the middle part of its course. The direct proximity of the lung and pleura indicates clearly enough the danger of ligature in this situation, and all cases hitherto undertaken have been attended with unfortunate results.

In front of the subclavian artery on either side is the common carotid, and between these vessels is the trunk of the vertebral and deep cervical veins; and in the middle line is the long cardiac nerve.

The vertebral vein is subject to many variations. Although in the vertebral canal it is generally a single trunk, it may form a plexus. In rare instances it joins with the deep cervical vein, and passing down behind the articular process, forms a trunk which receives the blood from the sinuses in the canal. It has also many variations in its point of termination, the most frequent of which is in the commencement of the innominate vein, and it may pass down hence either in front of or behind the subclavian artery.

In one case, on the left side of the body, it was found as a trunk in the vertebral canal, in front of the vertebral artery, and at its point of exit from the canal was directed forwards, and passed over the subclavian artery in front in order to terminate in the left innominate vein near the junction of the internal jugular. Thus it formed with this large trunk, on the inner side of the vertebral artery, a V, in which lay the thoracic duct before emptying into the subclavian vein.*

In a second case the vertebral vein came forwards from behind the subclavian artery, between it and the pleura, and terminated in the lower end of the internal jugular, so that after the removal of the pleura the vessel could be seen lying free, and crossing the subclavian artery from behind forwards. Into the horizontal portion of the vertebral, a vein opened corresponding to the deep cervical, in front of the subclavian artery. On the right side of the body, in a third case, it passed behind the subclavian artery, whilst on the left it passed in front of it,

* I have on more than one occasion observed the thoracic duct to terminate in the lower part of the vertebral vein.—TR.

and in a fourth case it passed down on both sides of the body in front of the subclavian artery.

These relations are important, inasmuch as in ligature of the ascending portion of the subclavian artery they are frequently met with, and care must be taken to avoid them. Directions are given for the avoidance of nerves and arteries, but no notice is taken of the vertebral vein, or of the thoracic duct, which on the left side lie close up. On the outer side of the carotid, immediately behind the sterno-cleido-mastoid, is the internal jugular vein, and between it and the carotid, the vagus nerve.

The external jugular vein is seen on the left side, between the clavicle and omo-hyoid muscle. On the right side it opens into the divided transversus scapulæ vein. The subclavian vein is not seen as yet, as it lies below the section.

In front of the trachea is the thyroid body, which is divided directly through its isthmus. It appears to be completely normal, both as regards structure and size, which in this country (Saxony) is seldom the case, as most subjects show enlargement of this gland.

The œsophagus at the level of the gland begins to leave the mesial line to get to the left side.

In Plate X, which gives the structures at the level of the sterno-clavicular articulations, the œsophagus already lies to the left side of the trachea. Although this lateral deviation of the œsophagus is the rule, still the exact level at which the greatest deflection takes place appears to vary. I find this lateral position complete in Pirogoff's atlas (tab. i, fasc ii), in which the section has passed between the first and second dorsal vertebræ, as in Plate VIII, where the œsophagus first begins to leave the middle line.

The head of the left humerus is divided nearly in its middle, and in front is part of the greater tuberosity, into which is inserted the tendon of the infra-spinatus. Under this tendon, and close to its insertion, is the thinnest part of the capsular ligament. The supra-spinatus, the mass of which is seen between the two bony ridges which belong to the scapula, is divided at its anterior extremity, at the point

PLATE VIII 73

where it ascends to its insertion into the greater tuberosity. Its tendon is blended with the fibrous structures on the anterior surface of the articulation.

The deltoid with its intermuscular septa is well developed, and between it and the insertions of the muscles attached to the greater tuberosity is a bursa, the cavity of which is indicated by a black line.

As the glenoid cavity has been divided nearly in the middle, the tendon of the long head of the biceps lies free in the joint. Beneath it was found a thin fold of synovial membrane, but higher up the tendon was completely free. · On the anterior surface of the coracoid process are the tendinous origins of the biceps and coraco-brachialis, and internal to them the fleshy mass of the pectoralis minor. On the posterior and inner side of this process the conoid and trapezoid ligaments are seen in section. The head of the right humerus is divided considerably higher than that of the left, namely, at the level of the upper border of the glenoid cavity. In consequence of this the articular cartilage appears completely encrusting the bone. The capsule is free all round, and the tendon of the long head of the biceps is seen coming up to be incorporated with the glenoid ligament.

Too much must not be expected from the plate, as the bundles of the tendinous masses can be only represented in general. The individual fibres of the tendon of the infra-spinatus, for instance, cannot be followed out round the head of the humerus. They become lost deeper down on the greater tuberosity, and are intimately blended with the insertion of the supra-spinatus.

If the section in this plate be compared with Plate IX in the large coloured atlas (also the section of a young powerful man) as well as with that of a man fifty years of age, its massive mould would be evident.

The individual layers of muscle are everywhere broader, although the skeleton itself does not appear larger or stronger.

The difference, therefore, between the longitudinal and horizontal measurements does not show itself in the manner which one would be led to expect from a superficial examination. For though the lower outline was drawn exactly to the section (and therefore closely corresponds with the plane in Plate IX in the large atlas), the breadth of the shoulder is

10

nearly an inch more than in the old man, whereas the antero-posterior diameter is half an inch more in the old than in the young man. Plate IX in the large atlas should not be incorporated with the series of the plates, as in the old man more abnormalities exist. There was a considerable enlargement of the liver, and a very great development of the thyroid body, so that the relations of the parts in the neck (as seen in Plate XXV) are much altered. The thyroid gland was enlarged below and on the left side, so as to encroach on a portion of the superior aperture of the thorax. It also pressed the left subclavian artery inwards and backwards upon the cupola of the lung ; the œsophagus also was pushed out of its place against the trachea, embedding itself between it and the vertebral column. The relation of the carotid artery to the sixth cervical vertebra has been already described, and it has been stated that the position of the arteries is to be defined not by the bone, but by the directions of the muscles and fasciæ, and that the bony prominences alone should not be considered of value as landmarks for finding the arteries. The same remark applies to the veins, nerves, trachea, and œsophagus in the region of the neck. These structures are so freely movable in the anterior region of the neck that the movements of the trunk or the pressure of a tumour may materially alter their position. This is particularly evident in Plate IX in the larger coloured atlas ; it can be estimated also in the present plate.

Such a change of position with regard to the skeleton is owing to the presence of the loose cellular tissue which envelopes these structures. But the relation of these important structures with regard to the muscles and fasciæ is constant, and consequently if an operation such as tracheotomy, œsophagotomy, or the extirpation of a tumour, has to be performed, the surgeon must make himself well acquainted with the fasciæ and muscles.

Tab. IX.

PLATE IX

THIS plate represents a section between the regions of the neck and thorax of a man twenty-two years of age, powerfully built and perfectly normal.

All the sections, as far as to that of the pelvis, in the following series have been made from this subject.

The lamina in this plate was about 1·4 inch thick, and its upper surface is shown, so that the body is viewed from above. The arteries were injected. The section passed just below the upper border of the manubrium sterni, and through the upper margin of the third dorsal vertebra, together with a small portion of its underlying cartilage.

Close to the sternum lie the clavicles in section and their interarticular fibro-cartilages. Laterally, near the sternal ends of the clavicles, are the sections of the first ribs, behind them those of the second, and further backwards and inwards those of the third. These last are not quite symmetrically divided, in consequence of the somewhat higher level of the right side of the chest. On the left side the third rib exhibits in connection with it a portion of the transverse process of the vertebra which articulates with it; whilst on the right side merely a small portion of the head of the rib is shown.

The scapulæ are divided through the glenoid cavities. The heads of the humeri show the greater tuberosities and the lower portion of their articular surfaces.

. If the plate which in the large coloured atlas is figured IX be compared with this (the position of the parts in the man of fifty years with that of a man of twenty-two years), it will be seen that in the former

instance the under surface of the third dorsal vertebra is divided, and in
the latter the upper. In the younger subject, also, the plane of section has
passed nearly an entire vertebra higher, yet in spite of this the sternum is
deeper.

It would follow, then, that although the section was perfectly horizontal the sternum lies higher in the young subject than in the old.
If the transverse and antero-posterior diameters of both the sections be
compared, the latter diameter is seen to be the larger in the old man,
whilst the transverse diameter is less. It is possible that the enlarged
thyroid gland invading the cavity of the thorax may be the cause of this
difference.

By comparison of the shoulders, we observe that in consequence
of the extremely powerful muscular development they stand much higher
in the young man; therefore considerably more is removed by the saw.
A glance at the large surface which involves the pectoralis major, the
deltoid, and the subscapularis, suffices to show that the muscular development has been great.

The anterior contour is partly owing to these masses of muscle and
partly to the different attitude of the shoulders. It may be that the older
subject was frozen with the arms slightly raised, and with the shoulders
pushed somewhat backwards, whereas in the present instance the arms
were laid close against the thorax.

By measuring the sections it is shown that the antero-posterior and
transverse diameters of the body differ but little.

The transverse diameter in this plate amounts to nearly an inch more
than in Plate IX in the coloured atlas, if both sections be reduced to
equal scale, since the bony contours are more regular. In the middle
portion of this plate the position of the vessels and nerves is more intel-
ligible than on the section of the body of the older individual, owing to the
changes in the relative position of the parts from the presence of the goitre
in the other case.

Behind the sternum lie the sections of the sterno-thyroids, and near
them and behind the clavicles are the sterno-hyoids. In front of the
sternum are the tendinous origins of the sterno-cleido-mastoids. Further

PLATE IX 77

in behind the muscle, the sternum, and the clavicles, is the upper portion of the thyroid body. It is separated from the sterno-mastoid by the middle layer of the cervical fascia. Further back is the left innominate vein, which, on account of its oblique course downwards from left to right, is extensively divided. The trunk can be followed towards the right side as far as the lumen of a vein, which is the inferior thyroid vein opening vertically. On the other side of this vein the trunk lies more deeply and is no longer seen through the cellular tissue. The isolated right innominate vein is divided transversely.

On examining the left innominate vein two small openings are seen; the anterior of these is the internal mammary vein, and the posterior the thoracic duct. The duct in this instance opens more internally than is usual, and into the innominate instead of into the subclavian vein; and it may be followed on the inner pleural surface of the left lung directly backwards, whence it bends downwards, applying itself along the vertebral column.

On the right side behind the great vein are a series of four large arteries, which pass obliquely towards the middle line. Commencing from the left side they are, the left subclavian, the left vertebral (which in this case sprang independently from the aortic arch), the left carotid, and the innominate. The arch of the aorta is immediately below the plane of section.

Although the length and point of origin of the innominate artery are liable to considerable variations, the vessel, nevertheless, lies so close to the middle line that it should be searched for in the middle line of the jugulo-tracheal space, as Pirogoff recommends. After researches on the dead body, I have convinced myself that this proceeding is the surest guide to the vessel.

The head is to be drawn towards the left side and the right shoulder depressed, and the tissues divided as far as the group of muscles coming from the hyoid bone and larynx. It is a matter of importance to make the incision exactly in the middle line between both sterno-thyroid muscles, and to divide the dense cervical fascia to which the great veins are intimately attached. If this be done the trunk of the artery in the loose cellular tissue can be isolated from the trachea, and the ligature passed. The surgeon must remember that close to it is the left innominate vein, which runs obliquely across its trunk, and that on the right side of the trunk, as is shown in the plate, the vagus nerve passes down. The vagus in this case was met with below the point of origin of the recurrent laryngeal nerve ; it therefore lies further back than it does higher up in its course on the left side ; the recurrent laryngeal nerve is between the œsophagus and trachea, and the trunk of the vagus is in front of the subclavian artery.

From the position of the innominate artery it is clear that burrowing of pus in the mediastinum is likely to follow such an operation as its ligature, whilst the relative shortness of its trunk and the strong pressure in the arch of the aorta are serious obstacles in the way of the formation of a resisting thrombus. It is therefore obvious that however artistically the operation itself may be conducted, it will be followed by serious consequences. The position of this artery must be taken into consideration in the performance of tracheotomy below the thyroid body. The surgeon must be prepared to meet occasionally with an arterial trunk from the innominate* running obliquely over the trachea (as Lücke did in one case). The artery is the thyroidea ima.

Ligature of the first part of the subclavian artery and its dangers have already been alluded to. It must be remembered that even in its normal relations (as in the present instance) the left subclavian artery lies in a niche of pleura, and that it has not been pushed against the pleura by means of the enlarged thyroid gland. (Plate IX in the large atlas should be referred to.) It can be readily seen from the plate that swellings of the thyroid body may push the œsophagus out of position, and displace the trachea backwards. In the superior aperture of the thorax the œsophagus normally inclines to the left side, and attains its greatest deviation in the region of the second or third dorsal vertebra. I have observed exactly the same condition in another section made on a normal male subject.

---

* In a case in which I performed tracheotomy on a man, æt. 50, I found the innominate artery running obliquely across the trachea below the isthmus.—TR.

PLATE IX 79

Pirogoff (tab. i, fasc. ii), in a transverse section made between the first and second dorsal vertebræ in a powerful man, shows the œsophagus placed at the side of the trachea; and, indeed, unless the œsophagus be much dilated (as in the case from which Plate I was taken), it does not project towards the median line. This fact renders it evident that in the operation of œsophagotomy, if there be no tumour of the thyroid body of the left side, the œsophagus must be looked for on the left side of the trachea; and from the plate it is clear that the operation is similar to that of finding the left common carotid or vertebral arteries. The close relation of the recurrent laryngeal nerve is to be noticed.

Under the pectoral muscles, on the outside of the cavity of the thorax, are the brachial plexus and subclavian vein, and between them is the subclavian artery. If the pectoralis major be removed with the muscular branches of the acromio-thoracic artery, a thin fascia is met with which passes over the short head of the biceps, the coraco-brachialis and the pectoralis minor. It extends inwards as far as the sterno-clavicular articulation, and envelopes the subclavius muscle. The fascia then passes upwards along the first rib, at the line of junction with the sharp-edged coraco-clavicular fascia, and terminates in a sickle-shaped margin. An aperture is formed externally and above, resembling the saphenous opening in the thigh, which permits of the passage of the cephalic vein, the acromial axis, and the external anterior thoracic nerve. Below this is Mohrenheim's fascia and the section has so passed that the continuity of this fascia is not interrupted, but is shown by means of a white line. The fascia forms with the posterior lamina a sheath for the pectoralis major and coraco-brachialis, and constitutes at the same time the anterior layer of the sheath of the axillary vessels. Higher up it attaches the vein to the subclavius muscle and clavicle. Wounds of the vein at this spot may be attended by a dangerous entry of air into the heart.

The posterior layer of the sheath of the vessels is formed by the fascia of the serratus magnus and intercostal muscles; the external layer being derived from the fascia of the subscapularis muscle. The cavity of the

shoulder-joint is indicated by the black line which marks out the capsule;
the folds are well shown which facilitate rotation of the head of the
humerus.

The strengthening of the capsular ligament by the insertion of the
tendons of the subscapularis and teres minor are well seen; and on the
right side the bursa, which lies between the tendon of the subscapularis
and the capsular ligament, is indicated.

FIG. 1.

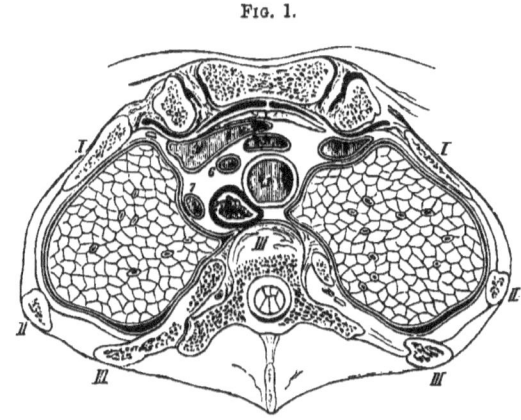

Subject A.—Male, æt. 22.   Normal.   Plate X.

1. Trachea.   2. Œsophagus.   3. Left innominate vein.   4. Right innominate vein.
5. Innominate artery.   6. Left common carotid artery.   7. Left subclavian artery.

In order to show by plane sections the changes in position which are
brought about by pathological conditions of the lungs and pleuræ, I have
arranged two of Pirogoff's plates so as to exhibit surfaces corresponding
with those in my own work, that is to say, viewed from above downwards.   They are reduced to half scale, and fig. 1 represents the central
portion of my own Plate X.

The surrounding muscles and upper extremity are not represented, in
order to make the woodcut clearer.

Fig. 2 represents large tubercular cavities in the upper lobe of the

PLATE IX 81

left lung, and is taken from a series illustrating dislocation of the heart and lungs.

Fig. 2.

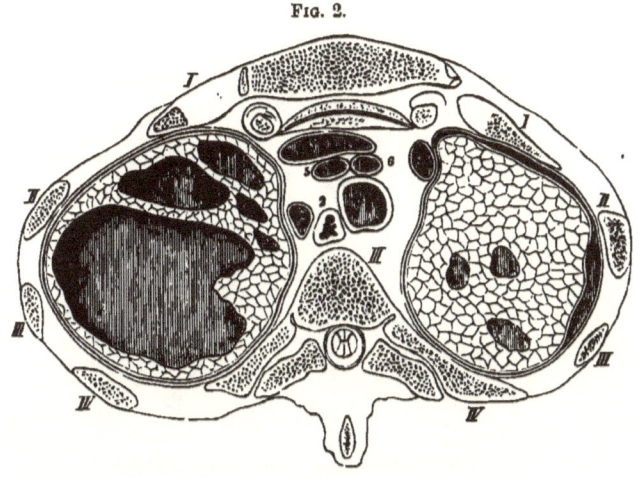

Subject B.—Male, æt. 18. Tuberculosis, Pirogoff, ii, 2, 3.

1. Trachea. 2. Œsophagus. 3. Left innominate vein. 4. Right innominate vein.
5. Innominate artery. 6. Left common carotid artery. 7. Left subclavian artery.

The section passes pretty much at the same level as in my own plate, and can therefore be conveniently compared with it. Pirogoff represents both portions of the section, as the saw had removed so much that there was considerable difference in the two sides. Pirogoff in his text, fasc. ii, p. 10, states that after freezing the body the upper extremities were removed with the scapulæ. The pulmonary and costal pleuræ were closely adherent. The cavities, which are shown by the deeper shading, have attained an enormous size, and the left side of the thorax was considerably more sunken in than the right; on the woodcut, however, it does not appear very remarkable. Between the first and second ribs only is a slight incurvation of the contour of the chest to be noticed. But the transverse diameter of the left portion of the thoracic cavity is considerably larger than the right. It is unfortunately not stated by Pirogoff whether any encysted pleuritic effusion existed lower down, which might have been the cause of this increase in breadth; consequently

11

there is little of importance to remark as to the cause of this altered form
of the mediastinal space.   Fig. 3 shows a section which corresponds with

FIG. 3.

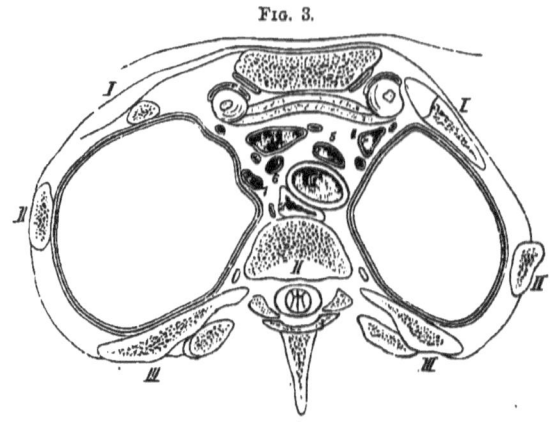

Subject C.—Adult male.   Left lateral pneumothorax.   Pirogoff, ii, 2, 3, ⅓.

1. Trachea.   2. Œsophagus.   3. Left innominate vein.   4. Right innominate vein.
5. Innominate artery.   6. Left common carotid artery.   7. Left subclavian artery.

mine, but the section has passed nearly a vertebra deeper.   It is from the
body of an adult male who, shortly before death, had pneumothorax of the
left side.

The apex of the left lung was so compressed by the mixture of pus and
air that it is not visible in this section; on the right side the lung was
divided near its cupola.   It is evident that the distension of the left side
of the thorax is not due to the elevation of the ribs only, but also to the
dragging inwards of the mediastinum; and in consequence of this the
structures in the upper portion of this thoracic cavity appear to be pushed
considerably out of their places.

Tab. X.

PLATE X

THE upper surface of the section is here shown; it is from the same body as the foregoing, and is about 1·4 inch thick, the saw through the inferior surface of the arch of the aorta, about one fifth of an inch below the division of the trachea, dividing the fourth dorsal vertebra just below its middle, and the sternum immediately below its articulation with the first rib, so that on the right side a small portion only of its cartilage is shown. The scapula is divided below its spine, and the humerus below the tuberosities. The section has passed just through the upper edge of the teres major, laying free the course of the posterior circumflex artery and a portion of the circumflex nerve. The nerve and artery pass directly into the deltoid muscle. The section shows clearly that both these structures must pass behind the humerus in order to attain the middle of the deltoid.

The axillary vessels and nerves lie on the subscapularis and under the coraco-brachialis. Their position with regard to each other is altered from the preceding section, the artery lying more between the nerve and vein, but so enclosed in the heads of the median nerve that it appears pushed from the vein by the great nervous mass.

The thorax is divided close to the lower border of the first rib, and on the right side of the sternum is a small portion of a costal cartilage, whose position corresponds with a broader section of the manubrium sterni than the preceding. Behind the sternum are the origins of the sterno-hyoid and sterno-thyroid; laterally are the intercostal muscles, which are attached to the second, third, and fourth ribs, and which help to close in the chest wall. The section of the fourth dorsal vertebra is seen at the back of the cavity, and is divided so close to its lower surface that the articular processes of the fifth dorsal vertebra come into the section; on the right side a small portion of the fifth rib is seen, and on the left side the boundary is formed by the intra-thoracic fascia.

The form of the section of the thorax is that of a heart as seen on a playing card, and is produced by the projection of the body of the vertebra and the recession of the hinder end of the ribs.   It has been remarked by Hyrtl (' Topog. Anat.,' 1860, i, 492) that this form is associated with the upright position of man, since by this formation the centre of gravity of the thoracic viscera is advanced nearer to the support of the trunk.   This advantage is not possessed by other animals, and one cannot maintain that this form is only a consequence of this upright condition, since, in the newly born infant, the curvature of the spine amounts almost to nothing (Pirogoff, *a a O*, fasc. i A, tab. xvi, fig. 3).   But this heart-shaped form of the section of the thorax exists in new-born children, as I can state from my own observations.   Pirogoff's transverse sections also show it (fasc. ii, tab. xx).   I find, however, that the relation of the breadth to the depth in children is considerably more variable than is that of the adult at a corresponding level.   In the newly born child the antero-posterior diameter is to the transverse diameter nearly in the proportion of 1 to 2, whereas in the present plate of an adult it is as 1 to 3 ; in the old man, on the other hand, the proportion is more like the child's, viz. 1 to 2·5.   The lungs are in the condition of expiration, and that to such an extent that during life the respiration pause was never reached.   As the contraction of the lungs after death is dependent on their elasticity, the size which they gradually assume must be so much the smaller the younger, sounder, and more elastic the said lungs are.   And as the contraction of the lungs depends proportionately on the position of the diaphragm, with the heart, liver, and spleen, there is naturally in young powerful individuals a higher position of the diaphragm and of its neighbouring organs after death than in the aged or diseased.   If the section of the old man be compared with the present plate, it will be seen that it is deeper by a vertebra (the sixth in the old man).   Consequently, in the definition of the position of the arch of the aorta, division of the bronchi, &c., the age of the individual must be always taken into consideration, and no fixed

PLATE X   85

level of a vertebra for the individual thoracic viscera can be given. The lungs themselves are divided through the lower portion of their upper lobes, so that on the left side a small portion of the under lobe falls into the section, which, as the plate shows, quickly increases in size downwards. Between the lungs, in front, is the thymus gland, which is sometimes found as late as the twentieth year, and consequently renders a mesial section possible on the young person, without opening the pleural cavity. In older subjects, after the atrophy of the thymus gland, both lungs lie so close to each other that in such a section it is impossible to avoid opening the pleural cavity.

I omitted to speak of the details of the form of the mediastinum; representations of it are given by Hyrtl, ' Top. Anat.,' i, 547, and by Luschka in Virchow's 'Archiv,' xv, 369.

There is nothing more variable in shape than the mediastinal space, for it is confined by fixed limits only in front and behind, on both sides the boundaries are moveable.

The alteration in capacity of the lung during breathing must also alter the position of the mediastinum. It further obtains that the contents of this mediastinal space are moveable and changeable. The œsophagus when distended takes up more space than when empty and collapsed. The same remark applies to the great vessels, which alter in size considerably after each contraction of the heart. The mediastinum in the region of the sterno-clavicular articulation, as shown in the plate, passes downwards and inwards, so that the space beneath it is contracted and funnel-shaped. In consequence of the position of the thymus gland it is possible to reach the upper edge of the arch of the aorta with its three branches, and the superior caval and innominate veins, without necessarily opening the pleura, and perforation of that portion of the trachea which lies behind the manubrium sterni may take place from the anterior wall of the chest without the pleura being involved. In order to compare the relations produced by pathological changes at similar levels, I have taken some reduced and reversed figures from Pirogoff, so that they may correspond, as far as the observer is concerned, with my own plates.

Fig. 1 is reduced from Plate XI of this Atlas.

Pirogoff's drawing, Fig. 2, which shows a body affected with left pneumothorax, was reversed and reduced so that it might be the more

Fig. 1.

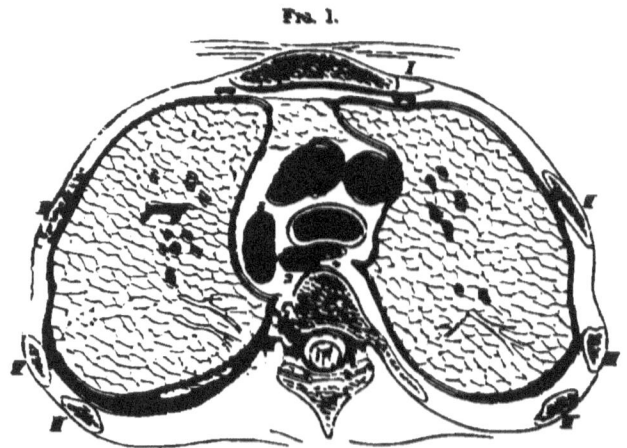

Subject A.—Thorax. Male, æt. 22. Normal. Plate XI. ½.

1. Trachea. 2. Œsophagus. 3. Superior vena cava. 4. Arch of aorta.

readily compared with mine. The section, according to Pirogoff's description, passed through the second intercostal space, and divided the third, fourth, and fifth ribs to the lower border of the second dorsal vertebra, so that in subject C the sternum must have been placed considerably higher than in my preparation. Whilst the posterior osseous portion shows relations similar to mine, the sections through the sternum differ by a rib and an intercostal space. This elevated position of the sternum can be readily explained from the pneumothorax, and the emphysema existing on the right side.

The left lung lies compressed upon the vertebræ by means of a pseudo-membranous cord which is attached to the wall of the chest; the right lung which is immensely distended by secondary emphysema, shows no folds in the pleura, such as are to be seen in my plates. The superior vena cava is compressed.

PLATE X   87

This thorax has an entirely different shape from fig. 1, being fully distended.   On account of the greater pressure in the left half of the thorax,

Fig. 2

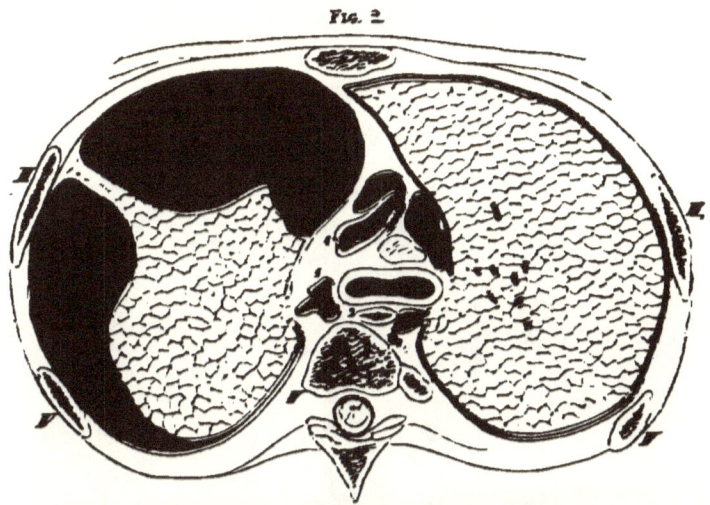

Subject C.—Adult thorax.   Left lateral pneumothorax.   Pirogoff, ii, b. 2, ½.
1. Trachea.   2. Œsophagus.   3. Superior cava.   4. Aorta.   5. Azygos vein.

the structures lying in the mediastinum, the trachea, œsophagus, and aorta, are pushed over towards the right side.   Pirogoff has figured more sections from each body, so that I was induced especially to indicate the individual subject with capitals, in order that the reader may be able to find the same body on the different sections.

Subject A is the powerful man from which my principal plates are taken.   Subjects B, C, D, &c., are from Pirogoff.   Subject C, according to his statement, is from a man of middle age who died in the hospital and had considerable pleuritic effusion.   I have found, moreover, a case of hydropericarditis with insufficiency of the semilunar valves of the aorta.

The section in fig. 3, which likewise is a reverse of a plate in Pirogoff's atlas, shows the same relations of the skeleton as mine.   The right lung, which was comparatively but little affected, corresponds almost exactly with mine. The anterior portion of the apex of the lung is slightly drawn over to the left side in consequence of the adhesion of the pleuræ to the remains of the thymus gland. The left lung shows important changes, due to infiltration and the formation of cavities. On account of

FIG. 3.

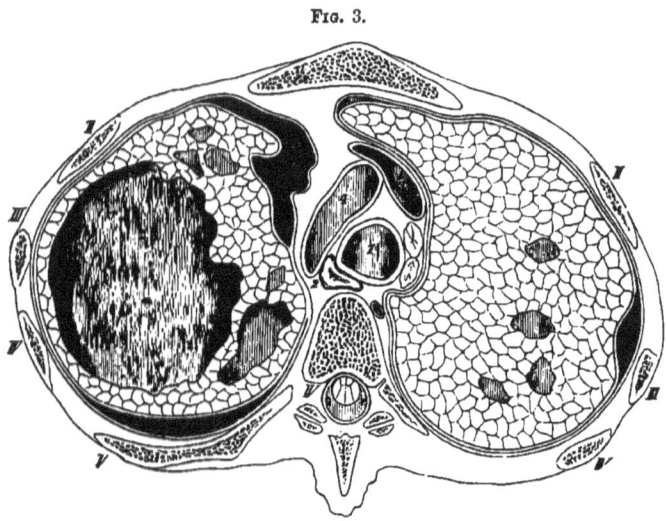

Subject D.—Thorax. Male, æt. 20. Tuberculosis. Pleurisy. Pirogoff, ii, 5, i, ⁴⁄₁.
1. Trachea. 2. Œsophagus. 3. Superior cava. 4. Arch of aorta.

the pleuritic effusion the left side of the thorax does not appear much sunken in. According to Pirogoff (p. 15, fasc. ii), the cellular tissue in the mediastinum was essentially altered by the previous inflammation. It shows strong attachments of the pleura to the surface of the ribs and to the inside of the mediastinal space, as well as adhesions of organs lying near to each other—a condition which cannot be intelligibly represented in the plate. The patient was a young man, æt. 20, who died in the hospital.

The woodcut fig. 4 is also from Pirogoff, and was taken from the body of a man who died of " scorbutic pleuritis," with great effusion of blood and pus in the pleural cavities. The anterior surface of the left lung was

PLATE X   89

so adherent to the thickened pleura, that the pleural cavity was divided into two portions, each holding a considerable quantity of blood and pus.

Fig. 4.

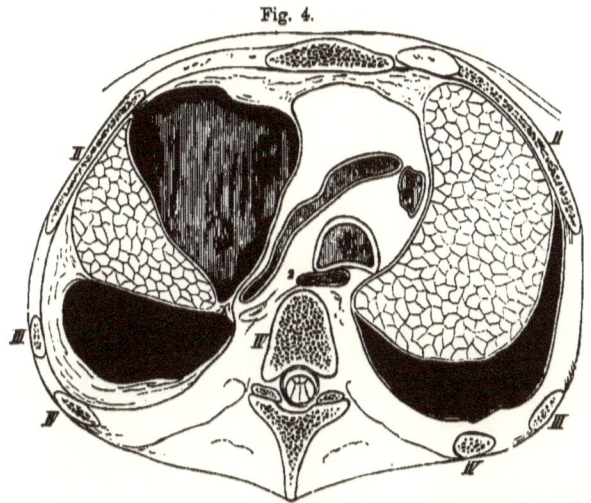

Subject E.—Male thorax. Lateral empyema of left side. Accumulation of serum in right pleural cavity. Pirogoff, ii, 18, 1, ½.

1. Trachea. 2. Œsophagus. 3. Superior cava. 4. Arch of aorta.

The left lung was adherent to the wall of the chest, and in consequence of the pressure from the pleuritic effusion the pericardium had become irregular in shape. The left side of the heart was much hypertrophied, and the mitral valve was covered with vegetations. The section which was made at the same level as mine, namely, through the middle of the first intercostal space, passed through the second, third and fourth ribs and divided the fourth dorsal vertebra in its lower half; great deviation of the mediastinum is shown. On account of the collection of fluid in the left pleural cavity the trachea is pushed over towards the right side, and the œsophagus lies the breadth of half a vertebra from its usual position towards the right side, so that deglutition must have been considerably interfered with. The plate also shows a dislocation of the superior vena cava almost to the middle of the right half of the thorax. In consequence

12

of the previous inflammation in the mediastinum a considerable amount of adhesion of the structures contained in it has been produced, whilst the arch of the aorta has been so dislocated, and its lumen so altered, that it appears as a narrow cleft. Such changes must have exerted their influence upon the heart; unfortunately they are not explained in Pirogoff's text. The change in position of the right lung was probably brought about by the organisation of the pleuritic effusion, especially noticeable in the sinking-in of the left half of the thorax, as seen at about the section of the second rib.

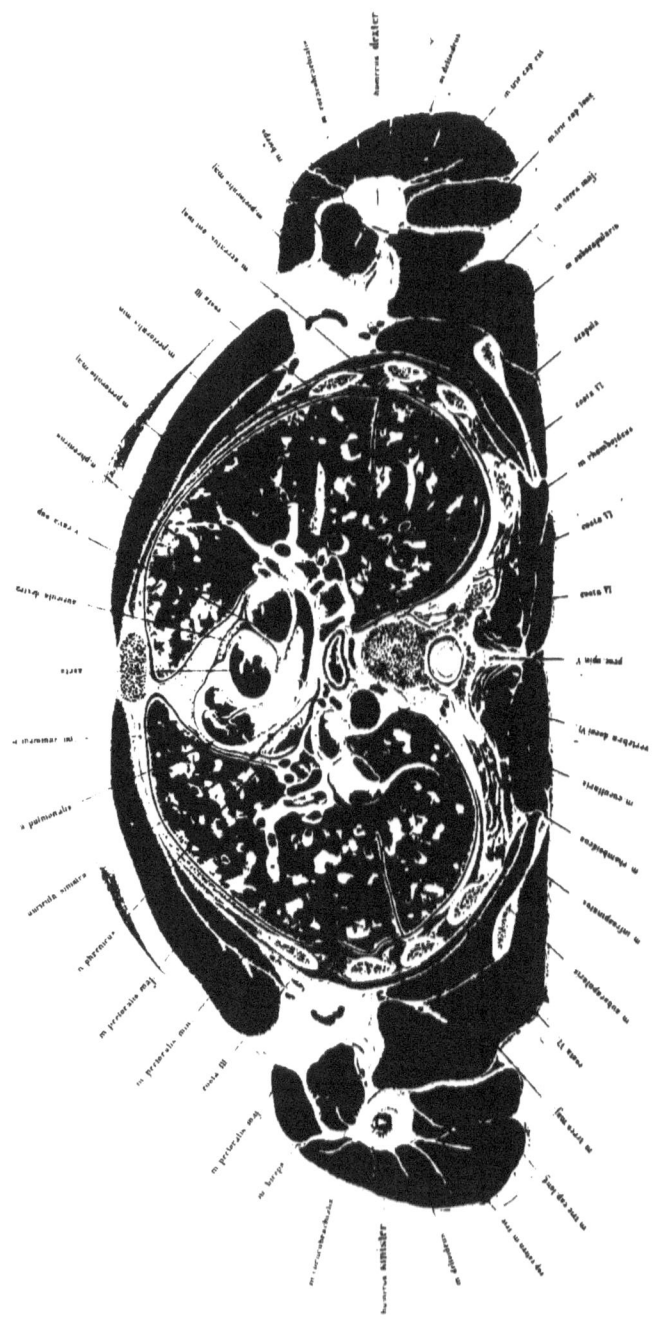

PLATE XI

THIS plate represents the upper surface of a lamina about an inch and a half thick, which was cut by a section passing through the trunk immediately beneath the sternal end of the second rib and the upper border of the sixth dorsal vertebra ; the saw passed out through the fat of the axilla, dividing the humerus at the insertion of the teres major.

Attached to the bone are the tendinous insertions of the great pectoral muscles, and on account of the position of the arms as regards the trunk they are so disposed as to exhibit a flat upward curve, and have been twice cut. Under the tendon of the pectoralis major lie the biceps and coraco-brachialis, and close under the last-named muscle are the vessels and nerves. The axillary artery is surrounded by the plexus, and is found next the muscle. The fascia of the coraco-brachialis must be divided to ligature this artery (after the arm has been raised), and the vessel should be reached from the sheath of the muscle, which can be easily drawn outwards; thus there will be little risk of pinching up and wounding the nerves and veins. Those portions of the trunk which are divided in the second intercostal space are of very great importance. The section of the great vessels passes immediately over their valves, and the left auricle with the upper wall of the auricular appendix are shown. The left auricular appendix lies more deeply, and is seen in front of the ascending aorta.

Immediately behind the sternum the lungs and pleuræ approximate each other so closely, that only a very small interspace remains. This narrow space leads from the anterior mediastinum to the region of the thymus gland; a sagittal section in the mesial plane in this body must have opened the right pleural cavity.

The contour of the pericardium is clearly shown; it extends at this level considerably further back on the left side than on the right, corresponding with the higher position of the left auricular appendix. On the right side it has been opened in front of the superior vena cava, and extends between it and the aorta posteriorly to the right branch of the pulmonary artery, playing the part of a bursa by permitting the necessary movement of these vessels upon each other. As the trunks of the vessels which pass from the lung into the left auricle, and from the right ventricle to the lung, run horizontally, a section which passes through the roots of the lungs exposes much more of their length, whilst the vessels of the greater circulation, which pass more vertically to and from the heart, appear divided more transversely. The pulmonary artery is the most important to examine of the vessels of the lesser circulation, as it is exposed for a large portion of its course. It is met with close to its origin, and its right branch is divided throughout its course. The left branch does not lie in the same plane, but it is also divided. It rises up somewhat in its course to the left lung, arching over the left bronchus and left auricle. The trunk of the pulmonary artery runs somewhat to the left, backwards and upwards, and can be seen in the upper surface of the section.

It is evident that the aorta and pulmonary artery are fixed together, whilst the former is capable of movement upon the vena cava. The relation of the aorta to the right pulmonary artery is important, as in aneurismal dilatations of the first part of the aorta compression of the right pulmonary artery may be expected. The position of the valves of the pulmonary artery and aorta with regard to the chest wall were accurately defined in the preparation, and can be deduced approximately from the plate. The pulmonary orifice lay behind the left border of the sternum under the upper margin of the third costal cartilage. The aortic orifice lay behind the left half of the sternum on a level with the third costal cartilage, behind and to the right of the pulmonary opening. The curvature of the aorta behind the first part of the pulmonary artery, with the position of its valves, is rendered as accurately as possible. It must however be expressly understood that such definitions cannot represent

PLATE XI 93

with absolute accuracy the relations on the living body. Apart from the influence which maintains the filling of the vessels, the position of the heart and its great branches is determined chiefly by the lungs and diaphragm, and it varies with each change of position of these important connexions. (This will be referred to again in the text accompanying the next plate.) Both bronchi are clearly seen; the left is divided more obliquely, in consequence of its being less vertical than the right, and as its ramifications lie in the plane of section, more of its branches are seen; whilst as the right has been divided more transversely, most of its branches have been separated. Between them, at the root of the lung, are a number of the characteristic pigmented bronchial glands.

Nearly in the centre, in front of the sixth dorsal vertebra, is the œsophagus, and behind it, to the left side, is the descending aorta, which already begins to take a direction towards the middle line. Between the œsophagus and the aorta is the thoracic duct, which in this instance is double. The vagus lies on the right side near the œsophagus and the vena azygos major; on the left side it lies between the bronchus and descending aorta.

The practical physician will notice with interest the changes which pathological conditions have given rise to in these sections. I have, therefore, introduced two plates from Pirogoff, which have been taken from the bodies of patients. Fig. 1, taken at the same level as my plate, shows extensive pericardial exudation.

The immense expansion which the pericardium has attained at the roots of the great blood-vessels is remarkable; both pleural sacs are widely drawn asunder, and the right especially has acquired a considerable inflexion. The pulmonary artery with its right branch has slightly changed its position with regard to the middle line; the aorta lies considerably further towards the right side than is normal, and is pushed far away from the vena cava. On account of the exudation all the vessels seem to be pushed towards the vertebral column.

The section corresponds with that given by Pirogoff (fasc. ii, p. 22), and passes through the second intercostal space, dividing the third, fourth, and fifth ribs of both sides and the fourth costal cartilage at the level of its upper margin. The age of the man, who had lain in hospital a long time, is not clearly defined, and is as of middle life. In any case the age was greater than that of my subject. It is remarkable that, although in the region of the sternum both sections began almost exactly at the same level, they struck different vertebræ; in Pirogoff's case the fourth, in mine the sixth. As the definition of the position of the heart with regard to the bones of the anterior wall of the

FIG. 1.

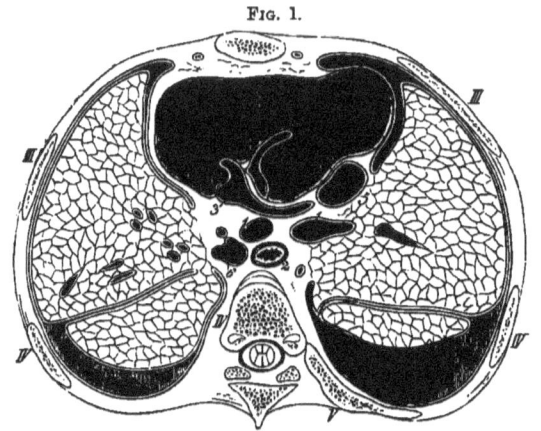

Adult male thorax. Hydro-pericarditis. Insufficiency of aortic valves. Pleurisy.
Pirogoff, ii, p. 1, ½.
1. Bronchi. 2. Œsophagus. 3. Pulmonary artery. 4. Ascending aorta.
5. Superior vena cava. 6. Descending aorta.

chest is of clinical importance, I have, in spite of the difference of the vertebræ of Pirogoff's plate, chosen it for the sake of the comparison, as the section happens to pass through the same intercostal space as mine.

It must be borne in mind that owing to the exudation into the pericardial and pleural cavities in Pirogoff's subject, the ribs are raised and their anterior extremities lie two vertebræ higher than in my case.

The following woodcut, Fig. 2, shows the variation in the position of the parts of a similar section in pleuritic effusion of the left side with pneumo-thorax. The subject is the same as in Fig. 2 of the text of

PLATE XI 95

Plate X. The expansion of the left side of the thorax, and the lateral displacement of the great vessels, can be clearly made out.

The commencement of the pulmonary artery lies behind the right

Fig. 2.

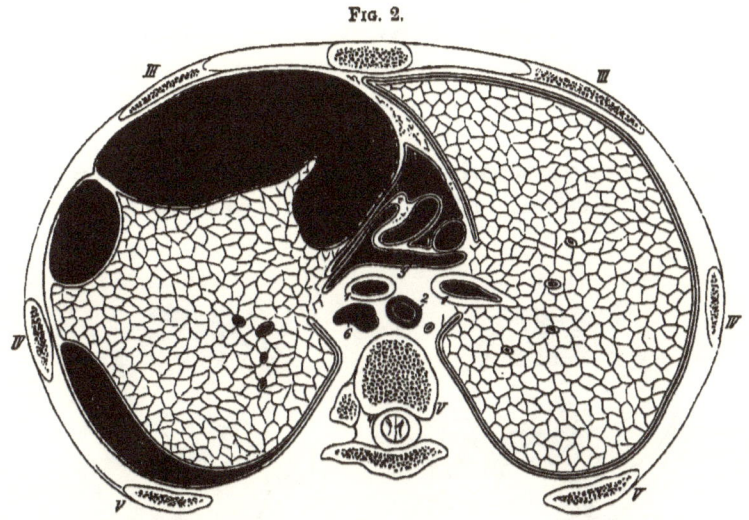

Subject C.—Adult male thorax. Left lateral pneumo-thorax. Pirogoff, ii, 7, 3, ¼.
1. Bronchi. 2. Œsophagus. 3. Pulmonary artery. 4. Ascending aorta.
5. Superior vena cava. 6. Descending aorta.

border of the sternum, and that of the aorta behind the third costal cartilage.

Pirogoff's section (cf. text to plate, 'Atlas,' fasc. ii, p. 28) runs horizontally through the upper border of the third costal cartilage, dividing the third, fourth, and fifth ribs of both sides; it passes also through the upper border of the fifth dorsal vertebra. Here the section passes a vertebra higher than in my case; this resulted from the expansion of the thorax, and the position of the ribs consequent on inspiration. It is remarkable that both right and left sides of the thorax are equally raised, so that they show a closely symmetrical division of the ribs. Besides the local pleuritic adhesion, which stretches like a cord from the inner surface of the ribs to the lung, there are other and wider bands which divide the pleural cavity into three portions. The left lung is, moreover, very much compressed and adherent to the costal pleura, so that its section appears polygonal.

As the normal relations of the thoracic organs have been exhaustively treated of by Luschka, Henle, Meyer, and others, I must refer the reader to their works, and proceed with a description of certain results of observations on dislocation of the heart from collections of fluid in the pleural cavities.

Fig. 3 shows the normal relations of the heart to the anterior wall of the chest, as determined from numerous examinations which I have

FIG. 3.

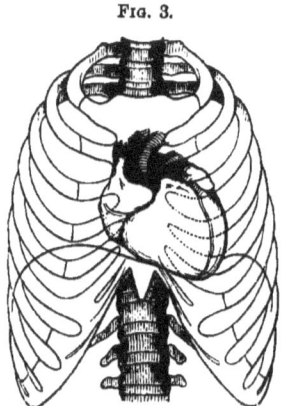

Normal position of the heart, ⅓.

FIG. 4.

Dislocation of the heart. Pleuritic exudation on the left side, ⅓.

made on young male subjects. After injecting the heart, and using only moderate pressure, the left auricular appendix became more visible than is usually the case when it is empty.

Fig. 4 represents a very considerable dislocation of the heart to the right side, produced by pleuritic effusion. The body lying on the back, the heart was fixed to the anterior and posterior walls of the thorax by six long needles, and the position of each portion accurately defined

PLATE XI 97

with regard to the anterior wall of the chest. It will be observed that the dislocation of the heart is considerably greater as regards its apex than its base, and that at the same time a rotation towards the right side on the long axis has taken place, so that a greater projection of the left ventricle has resulted. The vertical position of the heart's axis in this instance was determined by exact measurement.

The following woodcuts (5 and 6) also show dislocation of the heart from effusion into the pleural cavities. They are, however, the results of experiments which were made by myself on fresh normal bodies.

Fig. 5.

Fig. 6.

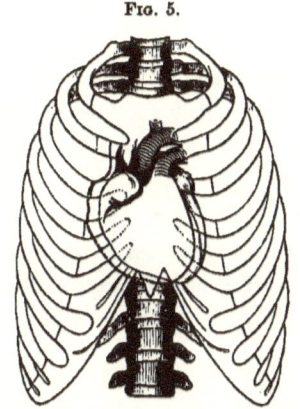

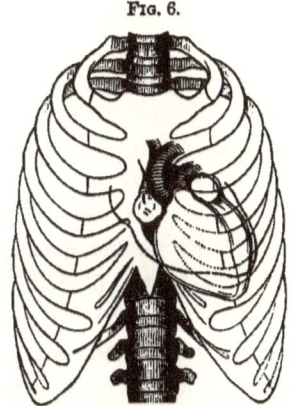

Left lateral hydrothorax, artificial, ⅛th. Right lateral hydrothorax, artificial, ⅛th.

The bodies were placed in the upright position and care was taken that the tracheæ remained open, and that the other parts were in their normal positions, and disregarding any experiment which did not seem to be complete, the conditions shown in the accompanying woodcuts were obtained.

After finishing the experiments by injecting a weak solution of common salt, the trachea was closed, so that on opening the thorax any farther falling together of the lungs should be impossible. The heart was fixed to the anterior and posterior walls of the thorax with long needles, and the intercostal spaces subsequently opened, in order to determine the position of the heart with regard to the framework of the chest. It was found that

13

the apex of the heart was pushed considerably backwards; and so also was the base, although strengthened by the great vessels forming the root of the lung. There was lateral rotation of the heart on its long axis. The quantity of fluid injected in fig. 5 was five, and in fig. 6 six pounds. It was observed that after the introduction of a pound and a half of fluid, there was an evident increase of dulness on percussion in the region of the liver (corresponding with the observations of Seitz and Zamminer).

As bearing on these experiments, I examined Pirogoff's plates relating to the sections of a subject with empyema of the right side and dislocation of the heart. (The section had been made after freezing.) I also collected material from the same author, of a body with pneumo-thorax of the left side.

After careful measurements on the different plates, the contours of the dislocated heart were constructed and shown in figs. 5 and 6 by the dotted lines, so that a comparison with the results of my own researches might be instituted.

In Pirogoff's definitions of the heart's position, exact as they are, the quantity of the morbid fluid could not have been measured, and one cannot expect that a dislocation of the heart could be expressed by a surface of the contours. Moreover, an artificial effusion into the pleural cavity could never produce the same relations as a gradually increasing exudation. But it follows certainly from these instances, and it is even proved by the difference of methods, that in dislocations such as these, the base of the heart does not remain fixed, but that it is considerably moved from its place (and the apex likewise), and that there is a rotation of the heart on its long axis.

Tab. XII.

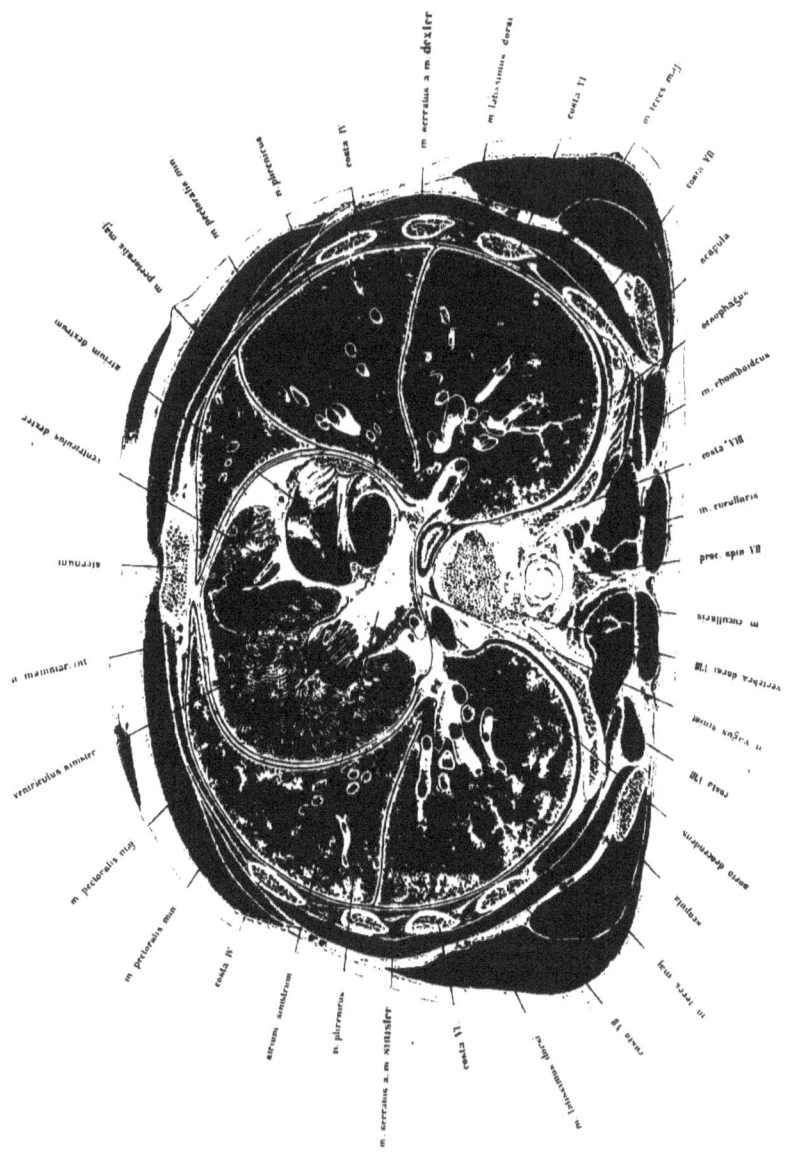

PLATE XII

THIS section, like the one just described, is viewed from above downwards, the thickness of the lamina being about one inch and a half. The section passed through both nipples and the third intercostal spaces, dividing the auricles of the heart and their valves. It passes backwards to the upper border of the eighth dorsal vertebra, and shows the eighth ribs of both sides; it cuts also the inferior angle of the shoulder-blade.

The great value of the plate consists in the fortunate section through the heart, both auricle and ventricle being opened. The left auriculoventricular opening is divided nearly in half, and the right is so cut at its upper border that a view is obtained of the ventricle. At first sight the cut surface of the heart and the space which this organ occupies seem immensely large, and yet a subsequent examination shows their relations to be normal. From the oblique position of the heart in the thorax a transverse section of the body would not divide it transversely but obliquely; therefore its walls appear much thicker than they really are.

The left auricle is divided not far from its base. The portion of it here represented shows a cavity about ·3 of an inch in its deepest part, while towards the right side the section rises to the level of the pulmonary veins. A small portion of the aortic segment of the mitral valve has been taken away; it will be found on the right side of the mitral opening.

Behind the left auricle the great cardiac vein is seen passing to the right auricle to open by the coronary sinus below the remains of the Eustachian valve. The point of opening lies too deeply to be clearly shown in the plate.

As the left ventricle lies more posteriorly and the right extends more anteriorly, the auricular septum is drawn out backwards and to the right side; the left auricle lies considerably higher than the right.

The inferior vena cava projects upwards into the posterior half of the right auricle, and in front of it are the remains of the Eustachian valve. Still more anteriorly the auricle bulges outwards and downwards to a depth of about an inch and a quarter, rising again to open into the right ventricle and by means of the auriculo-ventricular opening, which is guarded by the tricuspid valve. In front of the tricuspid valve is the right ventricle, which is opened by the section, and from which the section has carried away the root of the pulmonary artery. From the anterior wall of the ventricle (the section of which is seen in front) one of the musculi papillares passes backwards to the anterior flap of the valve, and behind this, deeper in the cavity of the ventricle, are the columnæ carneæ of the hinder wall. By comparison with the under surface of the next section the position of both auricles can be accurately determined. It appears that the cavity of the right auricle attains the level of the lower border of the fourth, to the middle of the third, costal cartilage, and that its corresponding auricular appendix reaches to the upper border of the third costal cartilage. Its greatest breadth extends from the middle of the left half of the sternum to about an inch external to the right border of that bone. The left auricle extends from the upper border of the fourth costal cartilage to the middle of the second intercostal space, and in breadth it corresponds to the eighth dorsal vertebra and its articulations with the heads of its ribs; its auricular appendix rises to the lower border of the second costal cartilage.

The right auriculo-ventricular opening is at the level of the eighth dorsal vertebra and to the right of the middle line of the sternum; it also extends across slightly to the left half of the body, nearly in the centre between the vertebra and sternum. Anteriorly its position is marked by the level of the nipple and the fourth costal cartilage.

The left auriculo-ventricular opening commences somewhat to the left of the sternum and reaches nearly to the middle line, lying 2·8 inches behind it at the level of the fourth intercostal space.

PLATE XII   101

A needle pushed into the middle of the third intercostal space, at the distance of rather less than half an inch from the left sternal border, would strike the central point of the mitral opening. In order to pierce the tricuspid opening, it must be thrust into the right half of the sternum at the level of its articulation with the fourth costal cartilage.

The pulmonary orifice would be reached at the upper border of the third costal cartilage, about one fifth of an inch external to the left edge of the sternum, and the aortic orifice at the level of the third costal cartilage.

I have frequently performed such experiments on young male subjects, and I am convinced of the accuracy of these statements. But I am far from insisting on their being absolute for all bodies, still less would I maintain that the positions are exactly the same for the living without further observation, entirely waving the question of pathological changes. According to the position of the body whether it lies on the back, side, or abdomen, so the position of the heart is affected, and further it is considerably influenced by the condition of the diaphragm. The heart is placed between the lungs and the diaphragm, so as to be surrounded by structures which can be displaced from it as soon as something else has taken their place. And owing to this arrangement the position of the heart is somewhat variable. The tender organ is not only perfectly protected from shocks which affect the anterior wall of the thorax, but has, moreover, free room for its own movements.

In the body of a young and powerful individual, such as the one here represented, the lungs gradually contract to an extent which is never the case during life. Consequently the external air presses equally on the surface of the abdomen and upon the diaphragm.

When the lungs contract, the heart, which lies between them, naturally moves upwards with the diaphragm, and so attains after death a higher level than is possible during life.

If the elasticity of the lungs be lost, as is the case in old people and in those affected with disease of the lung-tissue, we must expect a deeper position of the heart.

By sections in the bodies of young powerful men I found the pulmonary orifice at the upper border of the third left costal cartilage, and at the level of the sixth dorsal vertebra; in persons of from fifty to sixty years it lay below the fourth costal cartilage at the level of the eighth dorsal vertebra.

In the event of tympanites the inflated intestines push up the diaphragm and the heart until the latter lies between the yielding and more contracting lungs, so that the pulmonary orifice corresponds to the level of the second costal cartilage.

FIG. 7.

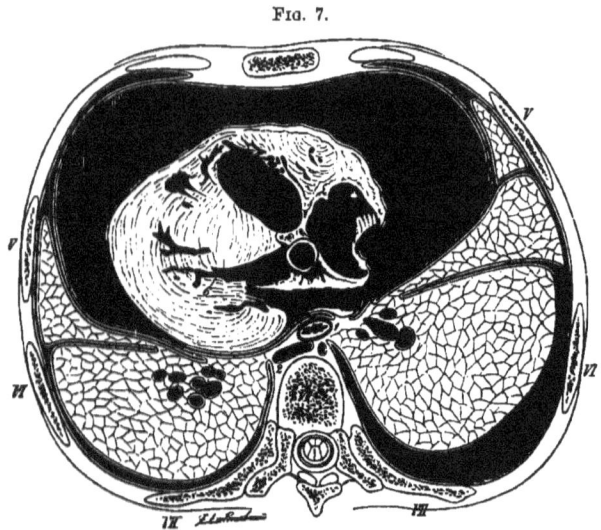

Adult male thorax. Hydro-pericarditis. Pirogoff, ii, 14, 4, ½.
1. Œsophagus. 2. Descending aorta. 3. Right auricle. 4. Left ventricle.
5. Left auricle. 6. Left ventricle.

The diameters of the chest have been discussed with Plates IX to XII, and the relation 1 : 3 has been tolerably well established. It will be seen that these relations are subject to essential changes in disease. For the purpose of comparison I reproduce two of Pirogoff's plates in woodcut.

The section, Fig. 1, is taken a vertebra higher than mine, consequently a small portion of the bulbus aortæ remains in front of the left auricle, of

PLATE XII   103

which a considerable amount is left. The aortic portion of the mitral valve is clearly seen lying stretched flat over the apex of the hinder flap. The right auricle exhibits in its posterior half the point of entrance of the superior vena cava, which has been somewhat compressed by the pericardial exudation, and in its anterior part is seen the entrance to the right ventricle.

If these relations be compared with the normal condition one is struck with the altered form of the thoracic cavity. The antero-posterior diameter is considerably enlarged; it amounts to the half of the transverse diameter, whereas it should be only one third.

Owing to the great distance of the sternum from the spinal column, space is permitted for the extensive exudation. The heart appears driven backwards, but this is not really the case, as the parts between the heart and vertebra, the œsophagus and descending aorta, have clearly ample room. But it is rolled over entirely to the left side.

The axis of the left side of the heart passes in a direction transverse to the section of the fifth rib, whereas normally it points obliquely forwards towards the left nipple. The axis of the right side of the heart shows a similar change in direction. The lungs are considerably compressed, to give more room for the pericardial exudation. Whilst in my plate they enclose the entire heart and closely approximate its anterior boundaries, they are here widely separated from each other and sunk back, notwithstanding that pleuritic effusion exists on the right side. The pleural cavities should be especially studied with reference to paracentesis pericardii, in consequence of their attachment to the chest-wall. In this section they are but slightly dislocated, only a small space near the sternum being left free, so that a trocar would have to be introduced very close to the border of the sternum in order to avoid wounding the pleura.

The section in Fig. 2 is taken almost exactly at the same level as mine, and the relations of the heart are similar, this organ being slightly pushed over, and at the same time rotated on its axis toward the left side. The left lung is considerably diminished, so that it is not applied to the anterior surface of the heart. The pleuræ, however, reach as far

as the sternum, a very small space existing between them; they exhibit
so many adhesions (according to Pirogoff's description) that the cavity
of the pleura was considerably interfered with.   In addition to the disloca-

FIG. 2.

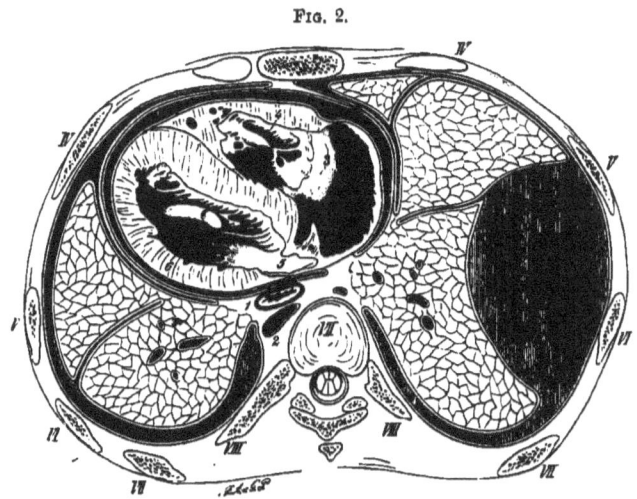

Adult male thorax.   Partial cystic empyema of the right side.   Pirogoff, ii, 11, 2, ¼.
1. Œsophagus.   2. Descending aorta.   3. Right auricle.   4. Left ventricle.
5. Left auricle.   6. Left ventricle.

tion of the heart, the remarkable pushing over of the œsophagus to the left
side is of interest; but unfortunately Pirogoff gives no further account
of this matter.

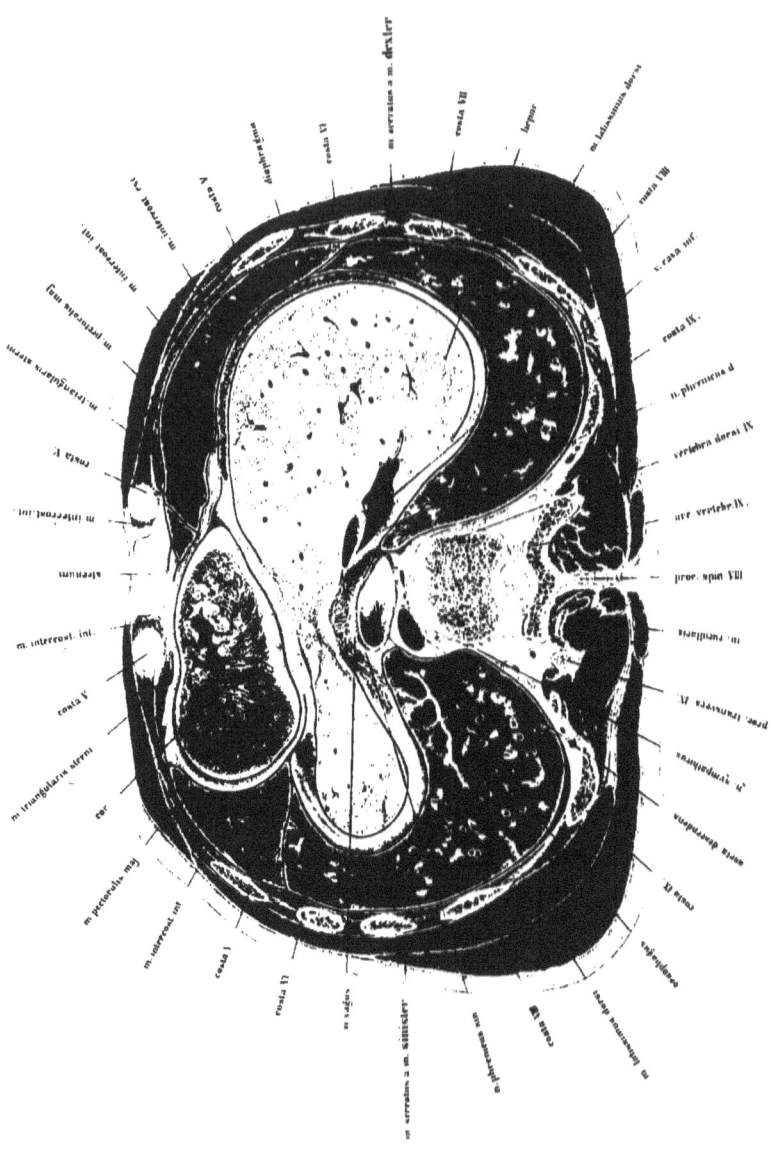

m. serratus a m. dexter

costa VII

hepar

m. latissimus dorsi

costa VI

A. cava

diaphragma

costa VII

m. intercost. int.

m. pectoralis maj.

m. triangularis sterni

costa V

m. intercost. int.

sternum

m. intercost. int.

costa V

m. triangularis sterni

cor

m. pectoralis maj.

m. intercost. int.

costa V

m. rectus

v. cava inf.

costa IX.

n. phrenicus d.

vertebra dorsal IX

arc vertebr. IX

proc. spin. VIII

m. medullaris

proc. transversa IX

n. sympathicus

aorta descendens

costa IX

oesophagus

m. intercostalis ext.

costa III

m. serratus a m. sinister

PLATE XIII

THE section of which the upper surface is here shown was taken two inches below the preceding; and passed through the lower portion of the sternum and the fifth costal cartilage; divided the apex of the heart, the diaphragm, and the liver; and came out posteriorly through the lower portion of the ninth dorsal vertebra, and the corresponding rib.

This plate terminates the series of sections of the thorax; and the abdominal cavity is already open, showing at a glance how wounds of the liver may involve the lung. Although the left lobe of the liver lies in the section, a very small portion only of the left half of the cupola of the diaphragm has been removed. It rises as high as the lower border of the fourth rib, seen from the front; whilst the right half, of which considerably more has been removed than of the left, rises as high as its upper border—nearly a rib's breadth higher, and almost on a level with the plane of the nipples.

It has been already stated in the last chapter that this position of the diaphragm does not correspond with its relations during life, but that it was so found in the body of a young powerful man, and that it would be pushed higher up in tympanitis.

The position of the heart is in immediate relation with the diaphragm and liver; and the lowest part of the heart is shown divided behind the fifth costal cartilage of the left side. The absolute apex of the heart is about four fifths of an inch from the plane of section. On the right side, in the apex of the right portion of the heart, is seen the lowest part of the cavity of the ventricle, filled with its columnæ carneæ. At the apex of the left side the section exhibits the arrangement of the muscular structure.

14

The heart does not extend downwards beyond the fifth rib, reaching only to its lower border; the cavity of the pericardium, however, extends about half an inch lower, and contains about a tablespoonful of frozen fluid. In a male fifty years of age I found at the level of the eleventh costal cartilage a portion of the heart corresponding with that here represented, but considerably deeper.

The relations of the pleuræ to the front of the heart are of practical importance. The pleuræ appear as folded sacs, which extend from the anterior border of the lungs towards the middle line, leaving in the present instance merely a small interspace between the left edge of the sternum and the fifth costal cartilage, through which the pericardium could be reached by the trocar without wounding the pleuræ. Bodies vary considerably in this particular, so that it is readily conceivable why so many different descriptions are given for the position of the point in the introduction of the trocar.

Luschka, however, is right when he maintains that the pericardium presents at the left border of the sternum a narrow strip quite free of pleura, so that it may be safely avoided in paracentesis of the pericardium. The safest method of operating, as I have satisfied myself, is to pass a fine trocar in the *upper* angle between the left edge of the sternum and the fifth costal cartilage. It does not appear justifiable to depend upon an adhesion of the pleura. Even large collections of fluid in the pericardium may exist for a considerable time without it.

The amount of extension of the liver towards the left appears surprising; hence the heart seems to be entirely supported by its left lobe, and from its abnormal size one is inclined to assume that some pathological condition was present. Such, however, was not the case, and the viscus was normal both in weight and structure.

It must be borne in mind that the left lobe of the liver shows great varieties of form even under normal relations; that it reaches down to the spleen; but that it lies always under the heart, a portion of which projects anteriorly and to the left side over the margin of the liver. Again, it is to be remembered that, in consequence, false notions are formed of the shape and position of the liver; one having been accustomed to observe it in front as

PLATE XIII 107

projected on a plane, in which case its entire extent cannot be shown. A good view of the extent and position of the liver is obtained from the diaphragm above; and this is the easiest method that can be adopted of studying the important relations of the liver to the spleen, stomach, and heart. I have frequently, after the removal of the chest-wall, shown the diaphragm intact, with a portion of the pericardium attached to it, and subsequently removed the diaphragm and introduced the liver into the drawing; and I always found a similar relation of the heart and liver to that seen in this plate, notwithstanding the variable extent of the left lobe. If the diaphragm be very carefully removed, the peritoneum may be preserved and the individual organs seen through it in their respective relations to each other. If the body be placed in the upright position, the pressure on the surface of the diaphragm is lessened and rupture of the peritoneal sac avoided. I give three plates which were made from the bodies of young powerful men (suicides) which were brought to the anatomical school with the rigor mortis on them.

There is no question that in such operations the position of the diaphragm frequently alters; and that with the removal of the upper half of the thorax especially the anterior and posterior walls of the lower half somewhat approach each other, and the cupola of the diaphragm rises correspondingly higher in consequence: this alteration of position having, however, but a very slight influence on the subjacent organs. A preparation of this kind may be made on a subject lying on the belly or on the back without any perceptible displacement of the enclosed viscera. Frequent observations show that by means of this method many useful results are obtained in explanation of the topography of this region. I have, then, rested satisfied with the representations obtained, and have refrained from attempting an improvement upon the plates by a previous moulding in plaster of Paris, and from using the drawing apparatus of Lucæ. Considering the sources of error which result from the relations in the dead body, an exact definition of the position of the parts must be given up.

Fig. 1 represents the relations of the parts, the stomach being tolerably full. This viscus when full pushes the left lobe of the liver outwards, and lies for the most part covered by it. The portion of the diaphragm that supports the pericardium indicates the position of the heart.

If the left ventricle, when full, exceeds the margin on the left side, it is clear that the heart lies, not on the stomach, but on the liver, and only its apex reaches the region of the stomach, and a transverse section would be similar to that represented on Plate XIII. The left cupola of the diaphragm is distended, therefore, by the left lobe of the liver, stomach, and spleen.

FIG. 1.

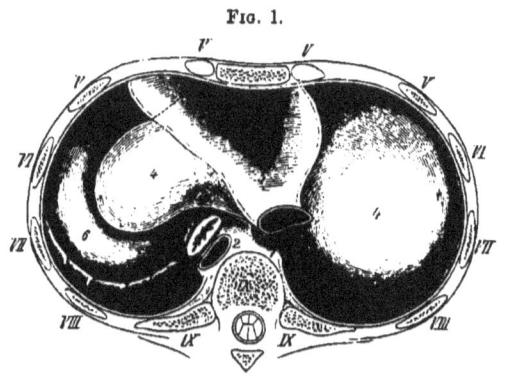

Normal position of the viscera below the diaphragm, viewed from above. ½.
1. Œsophagus. 2. Aorta. 3. Inferior vena cava. 4. Liver. 5. Pericardial portion of diaphragm. 6. Stomach. 7. Lobulus Spigelii. 8. Spleen.

Fig. 2 represents the position of the viscera below the diaphragm in still greater distension of the stomach. By simple inspection of the form of the circumference of the liver, it is evident that the figure was taken from another body, and that a body was used in which there was considerable distension of the stomach. This distension was not obtained by mere experiment, which very easily disturbs the relations of the parts : the subject was perfectly fresh, and the examination was made before it was touched in any way. The stomach, which was distended with food, did not extend as far as the left side, but still had against it the fatty portion of the peritoneum, which drags on the left end of the transverse colon, and which is continuous with the greater sac.

PLATE XIII   109

The left lobe has a different form from that in Fig. 1, notwithstanding that its relation to the heart is the same, or, at most, so slightly altered that the apex of the heart, in consequence of the greater breadth of the left lobe, has liver substance on the abdominal surface of the diaphragm under it. From observations that I instituted on different subjects, after filling the colon from the anus, or the stomach from the œsophagus, in order to demonstrate the variation in position of the organs in one and the same

FIG. 2.

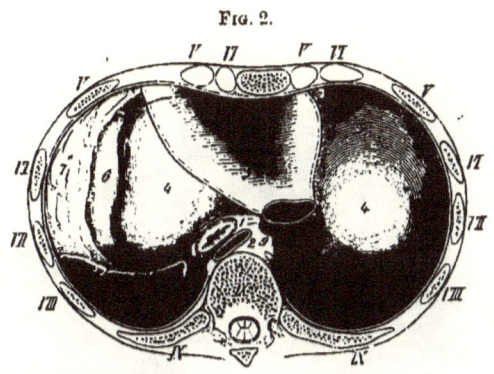

Normal position of the viscera below the diaphragm, viewed from above. ⅓.

1. Œsophagus. 2. Aorta. 3. Inferior vena cava. 4. Liver. 5. Pericardial portion of diaphragm. 6. Stomach. 7. Great omentum. 8. Spleen. 9. Lobulus Spigelii.

individual, I was convinced that even by carefully lifting the peritoneum, I obtained no condition of things from which a plate of any value could be made. The stomach was much displaced from its natural position, and was emptied with as much difficulty as the colon; so that I was forced either to use different subjects for the plate, or to select from them those which showed the organs in the state of distension desired. It appeared in the highest degree remarkable that in a portion of the trunk, to which merely the under half of the thorax was attached, one could inject a large quantity of water through the œsophagus, and leave it any length of time without its escaping. On introducing the finger through the œsophagus into the stomach one could feel its wall between the cardiac extremity and the fundus jutting out so sharply as to form a distinct valve. It must remain for further investigations how far these relations on the subject can be applied to the living body.

Fig. 3 shows the stomach empty, and the resulting space filled up on the left side by the colic flexure. The other relations are similar to the preceding. It appears in these plates that the heart always has the left lobe of the liver between it and the stomach, and lies on the stomach by only a portion of its apex, which may vary greatly in size. A

Fig. 3.

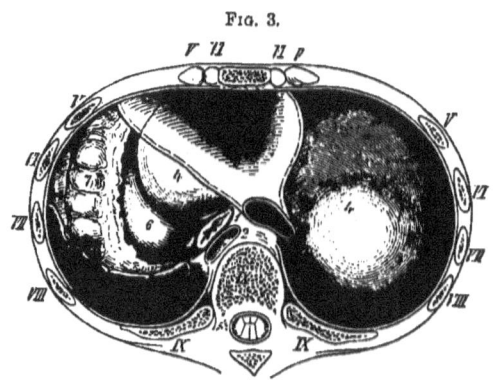

Normal position of the viscera below the diaphragm, viewed from above. ½.
1. Œsophagus. 2. Aorta. 3. Vena cava interior. 4. Liver. 5. Pericardial portion of
diaphragm. 6. Stomach. 7. Left flexure of colon. 8. Spleen. ½.

frontal section shows the same condition, and the order of these structures as arranged one above the other can be well studied (compare Henke, ' Atlas der Top. Anat.,' tab. xxxv, xxxvii, and Pirogoff, I A, ii A, ii B).

It will also be seen that, according to the condition of the stomach, the position of the viscera in the left cupola of the diaphragm will be altered. The left flexure of the colon is pushed up if filled with gas and the stomach empty; and will, as it more often contains air than the stomach, afford especially a full tympanitic percussion note in the lower half of the left

PLATE XIII   111

side of the thorax; it may also, by the strong pressure exerted upwards, disturb the functions of the organs within the chest.

The following woodcuts are taken from Pirogoff's atlas to demonstrate the change in the position of the apex of the heart as occasioned by pleuritic or pericardial exudation.

Fig. 4 illustrates the relations of the parts, at the same level, when the pericardium is very much distended with fluid. The section is taken at

FIG. 4.

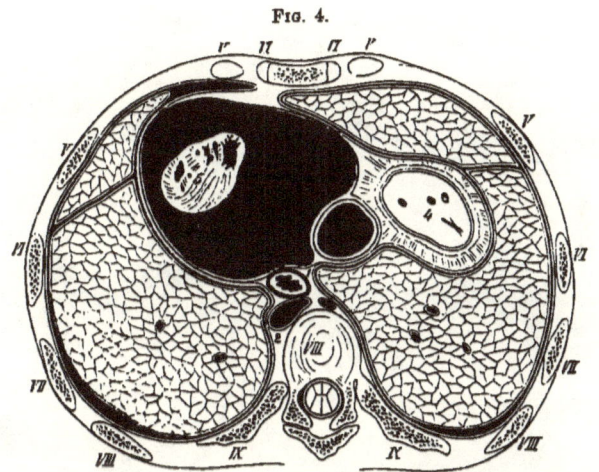

Male thorax. Hydro-pericarditis. Lungs healthy. Pirogoff, ii, 15, 2. ½.
1. Œsophagus. 2. Aorta. 3. Vena cava inferior. 4. Liver. 5. Heart.

the same level as mine, and the apex of the heart is pushed strongly backwards and somewhat to the left side.

The pleuræ approach each other in front, leaving only a narrow space at the left edge of the sternum. One would expect a greater separation of the pleuræ from each other as the quantity of fluid in the pericardium took up greater space. It is therefore the place to choose for puncture of the pericardium, as has been stated before, so as not to open the pleural cavity. Pirogoff does not mention the age of the individual; it is merely noticed that the lungs (and very likely the pleuræ) exhibited no abnormality.

Fig. 5 is a section showing the relations of the organs in pleurisy and hydropericarditis. It was made on the body of a man of middle age, who died in hospital, and passes deeper than my section by a vertebra. Notwithstanding the mass of exudation, very little of the liver is divided. As regards the position of the apex of the heart, it is dislocated backwards

FIG. 5.

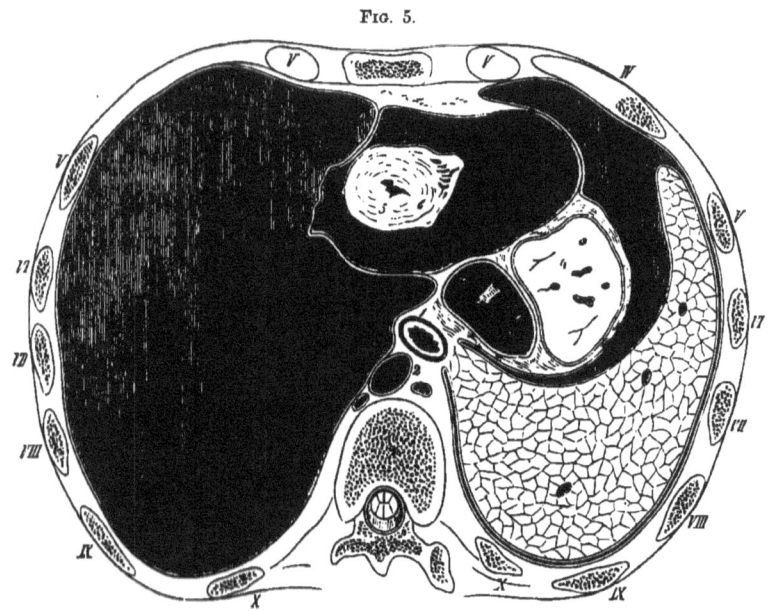

Male thorax. Left pleurisy. Hydro-pericarditis. Pirogoff, ii, 22, 2. ¼.
1. Œsophagus. 2. Aorta. 3. Vena cava inferior. 4. Liver. 5. Heart.

and to the right. The distension of the left pleura is so considerable that it extends forwards to the middle line and posteriorly beyond it.

Of the ribs of the left side almost the same are divided as in my case, from which it is evident that the effusion was more considerable, causing a tilting up of their anterior extremities. On the right side, on the other hand, which, according to Pirogoff's account, contained very little fluid, the ribs lie wider apart, so that the fourth rib is sawn through.

PLATE XIII   113

Fig. 6 shows the relations of the parts in double pleurisy and hydro-pericarditis. The description is to be found in Pirogoff's atlas, ii, p. 54.

The section, which has passed a vertebra deeper, divided the fifth, sixth, seventh, eighth, and ninth ribs of both sides, and shows almost the same relations of the skeleton as Plate XIII, both halves of the thorax being symmetrical. The man had an encysted empyema of the right side. The right lung was strongly compressed, and appeared polygonal in section

FIG. 6.

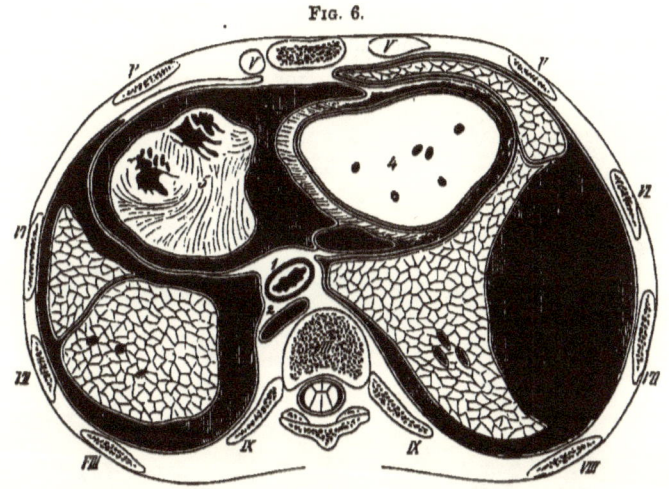

Male thorax. Partial cystic empyema of right side. Hydro-pericarditis. Pirogoff, ii, 15, 4. ½.
1. Œsophagus. 2. Aorta. 3. Vena cava inferior. 4. Liver. 5. Heart.

in consequence. The left pleura was thickened and very adherent. The heart, it will be observed, is dislocated and drawn to the left. The left lung lies far back, and its pleural sac is firmly adherent for its whole length in front of the heart, so that puncture of the pericardium could be performed without danger of the pleura at the sides. With regard to dislocation and hypertrophy of the heart, some authors have frequently observed a bending in of the inferior vena cava. (Compare Luschka, 'Anat.,' i, 2, p. 445; Bartels, 'Deutsches Archiv,' iv, p. 269.)

In my opinion the question is not yet decided, and can only be definitely

15

settled by allowing a body to be frozen, and to expose the right auricle with the venæ cavæ from behind with hammer and chisel. Transverse sections, like the one under observation, where the vena cava and the entrance of the hepatic vein are cut through immediately below the foramen quadratum, throw little light on the question; nor can much be expected from experiment or clinical observation. Researches on animals, which I have instituted in Ludwig's laboratory, and published in the reports of the Academy, show that ligature of the inferior vena cava does not set up any considerable disturbance of the circulation, as the blood finds a ready path collaterally by means of the azygos veins and spinal plexus, thus getting into the superior vena cava.

Tab. XIV.

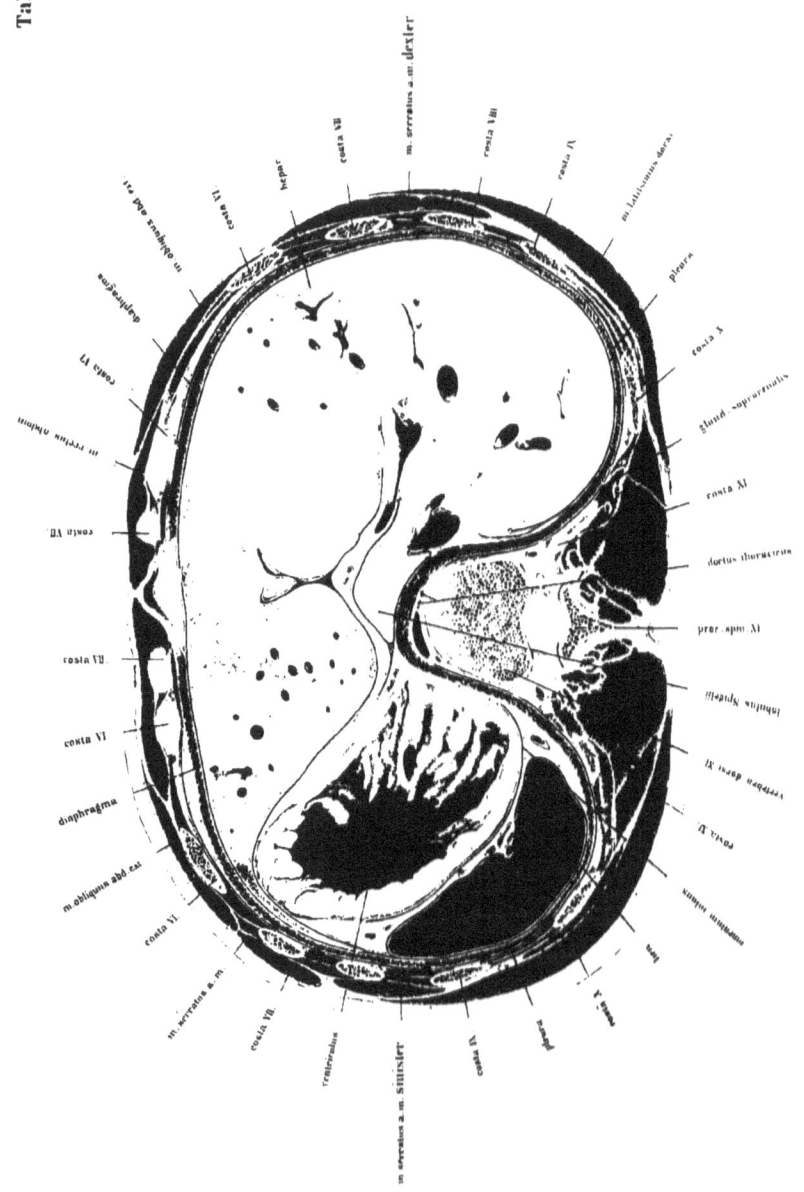

costa VII

hepar

m. serratus a. m. dexter

costa VIII

costa IX

m. latissimus dorsi

pleura

costa X

gl. suprarenalis

costa XI

ductus thoracicus

proc. spin. XI

lobulus Spigelii

vertebra dorsi XI

costa IX

systema nutrens

ren

pancreas

costa IX

m. serratus a. m. sinister

ventriculus

costa VII.

m. serratus a. m.

costa VI.

m. obliquus abd. ext.

diaphragma

costa VI

costa VII

costa VII

m. rectus abdom.

costa II

costa XI

m. obliquus abd. ext.

diaphragma

hepar

costa VIII

PLATE XIV

THIS plate represents a section through the epigastrium, exposing the liver, stomach, and spleen. No more is to be seen of the lungs; the black line immediately internal to the ribs represents the pleural cavity, whilst close to it is the diaphragm, appearing as a muscular ring. The structures lying external to it belong to the thorax, and internal to the diaphragm is the abdominal cavity.

The plate is taken from the upper surface of a section two inches thick from the same body as the preceding and the following.

The body of the eleventh dorsal vertebra is seen divided near its under surface, so that a small piece of the interarticular fibro-cartilage is shown. The arch lying behind it also belongs to the eleventh dorsal vertebra: the joint spaces in front belong to the articular processes of the twelfth: and on either side are the sections of the eleventh, tenth, ninth, eighth, seventh, and sixth ribs; the seventh and sixth ribs being divided twice, but not the xiphoid cartilage, since the section passes below it. It appears strange, at first sight, that the section of the right half of the body should have a larger area than the left, the transverse diameters differing by about half an inch; the cause of this is, however, in some measure, owing to a want of symmetry, and also to the fact that the saw-blade diverged somewhat from the horizontal plane.

The liver occupies the greatest amount of space, and is perfectly normal in structure and weight. The left lobe of the liver is prolonged into a thin lamina, which is stretched over the stomach almost as far as the spleen. This explains the great extent of the liver in the left cupola of the diaphragm, in the preceding plate. At the point of junction of the right and left lobes, in the left longitudinal fossa, is the ligamentum teres in a fold of peritoneum; and posteriorly, lying on the diaphragm, the

lobulus Spigelii with the omentum.   Close to it on the right is the inferior
vena cava;. and in front of this are the transverse fissure, the portal vein,
and the hepatic duct.

The stomach contained some frozen food, which was removed so as to
show its walls.   It was ascertained subsequently that the fundus of the
stomach had attained its highest position, and that beyond the distended
portion in the commencement of the stomach, there was contraction
where the folds of the mucous membrane were most marked, and that
subsequently the cavity became again distended further to the right side
and below.   It appears, therefore, that Luschka is quite correct in
disputing the entire approximation of the anterior and posterior walls of
the organ in its empty condition.   Here also, where completely normal
conditions existed, the stomach was contracted like intestine in its empty
portions, and was not flat, as represented in some plates.

The cut edges of peritoneum behind the stomach belong to its lesser
sac.   Further back is the spleen, normal, with its blood-vessels ; it corresponds with the course of the ninth, tenth, and eleventh ribs, and in
its greatest diameter follows their direction.   The left supra-renal capsule is not seen, whereas the right is evident between the liver and
diaphragm.

Concerning the relations of the peritoneum, it must be remarked that
transverse sections are not adapted for displaying it.   The cavities can
only be represented by black, and the cut edges by fine white lines, which
easily mislead the eye.

In order to render any representation advantageous, views of surfaces must be given, or longitudinal or oblique sections taken as the
bases of the drawings, by which, semi-diagramatically, the cavities of the
peritoneum appear enlarged.   Luschka and Henle have excellent plates of
this kind.

On the other hand, representations of such transverse sections, here
shown true to nature, are of great value surgically.   They show what
localities are free of peritoneum, and what not, and the surgeon consequently can plan an operation.   It is of the first importance to avoid this
membrane, and in this, as in the following plates, the boundaries of the

PLATE XIV   117

cavities and points of investment of the peritoneum have been represented with the greatest fidelity; and hence the cavity of the omentum between the lobulus Spigelii, and the posterior wall of the stomach, was not drawn as contracted, although both cavities are connected immediately below the surface of the section. The portion of the stomach which lies close to the diaphragm shows the end of the posterior cardiac region free, and uncovered by peritoneum.

An examination of the peritoneum shows that it has two functions to perform, especially mechanical, which are — (1) that it fixes the several organs in the abdominal cavity in definite places; and (2), like a colossal, sinuous, mucous membrane, allows of their movements upon one another in their various conditions of distension. These changes of position can occur, where the black lines in the plate, like joint spaces, represent the cavities of the peritoneum; at points, on the other hand, where the peritoneum is reflected, and leaves a free space for the entrance of blood-vessels, the viscera are fixed to their surroundings.

In order to show the relations as they exist in the extremes of age, I have here introduced two woodcuts. Fig. 1 is taken from a man, æt. 50, who had enlargement of the liver and spleen. Fig. 2 from a recent body of a female infant, at full period, born dead.

The body of the old man is the same which furnished Plate IX in the large coloured atlas. Death resulted from hanging, and the stomach and intestines were empty.

The section passed through the tenth dorsal vertebra, and anteriorly through the xiphoid cartilage. The stomach was empty, with the exception of a little frozen mucus. The lung structure was normal and absolutely empty of air. The liver large and fatty. Supra-renal capsules and spleen large.

The well-developed body of the child showed no irregularities.

The great resemblance between Figs. 1 and 2 is singularly remarkable, also the fatty livers of the old and young subjects; moreover, the relations correspond wonderfully accurately.

In both cases the liver fills up almost the whole interspace internal to the diaphragm, and spreads over a large portion of the spleen, which lies

in relation to the spinal column, as in Plate XIV.   The stomach alone
shows any important change of position.   In both cases it is empty;
has the same position between the left lobe of the liver and the spleen;
has a portion of the diaphragm for a covering; and is not overlaid by
peritoneum.   On the other hand the shape is different in the two subjects.

Fig. 1.

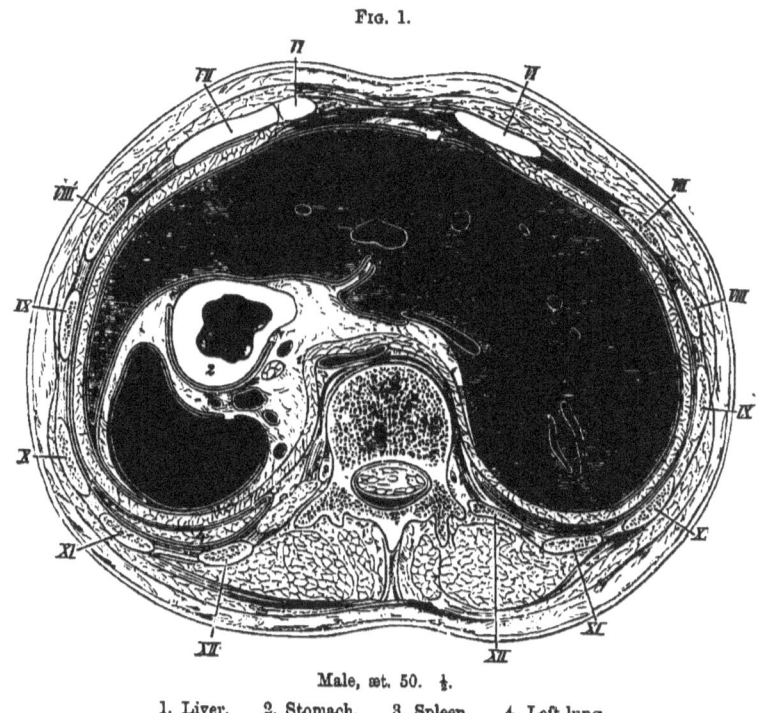

Male, æt. 50. ½.

1. Liver.   2. Stomach.   3. Spleen.   4. Left lung.

Whilst in the former case the stomach is contracted like intestine, in the
latter it is an oblique chink; so that the anterior wall lies relaxed on the
posterior—a condition I have never observed in the adult.

   In front of the right supra-renal capsule is, in the child, the inferior
cava, lodged somewhat deeper under the lobulus Spigelii than in the
other instance, where the capsules are not seen, notwithstanding that the

PLATE XIV   119

section has passed three vertebræ lower. On the other hand, corresponding with the slight power of contraction of the lungs in the old man, they
are still visible at the level of the first lumbar vertebra; whilst in the youth
of twenty-two years in Plate XIV the pleural cavities are empty at the
eleventh dorsal, and in the new-born child, Fig. 2, at the tenth.

In the new-born child the thorax is fixed at extreme expiration, to

FIG. 2.

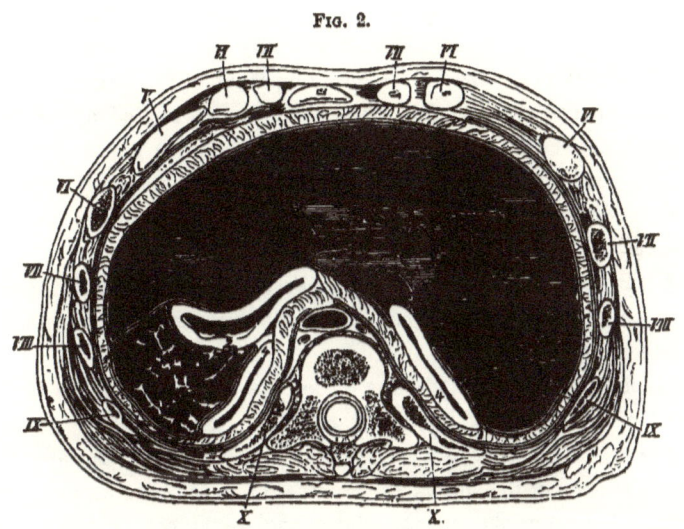

Child, at full period.   Born dead.   Natural size.

1. Liver.   2. Stomach.   3. Spleen.   4. Supra-renal capsule.

which it can never return after the first inspiration. The entire contents
of the upper portion of the abdomen must therefore be depressed as soon
as the diaphragm, during the first inspiration, leaves its high position; and
the figure, which here lies three vertebræ higher than in the old man, would
then take a considerably deeper level.

As in Plate XIV the space between the liver and the spleen appears
to be completely filled up by the stomach, which, however, presents only a
slight degree of distension, the question arises, what would be the condition
of things if this viscus were more distended? It is easily seen that, apart

from a considerable protrusion of the anterior wall of the abdomen, which
is observable after each full meal, the lower ribs also must give way—a
circumstance which, under a continued swelling of the abdomen, leads
even to permanent prominence of the thoracic segment, as may be proved
in many ways, and is especially seen in children. The left lobe of the liver
must follow more or less the movements of the stomach, since it forms a
species of covering to that organ; it is lifted up by the distended stomach,
pushing the pericardium up with it, and sinks down with the contracting
stomach, the place of which is taken partially by the left flexure of the
colon. The mesentery-like left coronary ligament of the liver renders possible such movements of its left lobe, which are associated either with a
turning of the entire liver (the axis of which is to be sought in the
right lobe, corresponding with the strong, firm attachments to the right
half of the diaphragm), or arise from the yielding or distension of the
soft tissues.

Fig. 3 reduced from Pirogoff will make this relation clear, even if one
does not obtain an entirely correct idea of the form and position of the left
lobe of the liver.

From this cut it is clear that the spleen lies so far back that any determination of its posterior limits by percussion is impossible. It is true
that by percussing in a horizontal direction around the thorax towards the
spine, at the level of the spleen, we obtain a different sound as we approach
the spine; but from the present plates one would not be warranted in
assuming the existence of an air-containing organ between the spleen and
the spine. We must look for the cause in the alteration of the elasticity
of the ribs at this point. Further, we always find, if we percuss in a
vertical direction on the back and in the axilla from above downwards,
that the commencement of the dulness is in a horizontal line, corresponding with the limit of the base of the lung, and covering the superior
portion of the spleen, which is directed obliquely downwards and forwards. One can easily convince oneself of the firm position of the spleen,
which is especially owing to the reflexion of the peritoneum, under the
name of phenico-splenic ligament, if the upper portion of the thorax
be removed on the dead body, and the sac of peritoneum preserved,

PLATE XIV   121

so that the liver, stomach, left flexure of the colon and upper wall
of the spleen are seen through it. The stomach can be inflated
and again allowed to collapse, and the descending colon filled and again
emptied, when it will be always found that the upper border of the spleen
is unchanged in position.

FIG. 3.

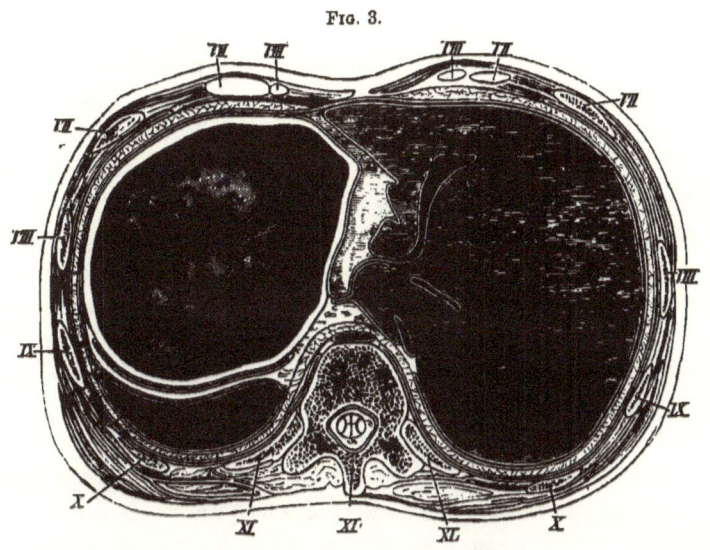

Youth, æt. 15.   Stomach distended with air.   Pirogoff, iii, 3, 1·½.

1. Liver.   2. Stomach.   3. Spleen.   4. Abdominal aorta.   5. Vena cava inferior.

On laying the subject on the abdomen the spleen does not sink
forwards, but remains in its original position. The relations are, however, different when the attachments of the spleen to the diaphragm are
sparse or easily lacerable, or drawn out into long bands. This might
account for the so-called movable spleen.

16

PLATE XV

In this plate the 1st lumbar vertebra is divided in the middle; on the right side are the sections of the twelfth, eleventh, tenth, ninth, eighth, seventh, seventh and eighth ribs, the seventh and eighth being twice cut; and anteriorly the arches of the cartilages appear. On the left side the twelfth rib is absent, as from being so short it is not met with in the section, but lies entirely in the preceding lamina. This section, like the preceding, exposes the upper portion of the abdomen, with a part of the spleen, stomach, and a large part of the liver. The diaphragm is divided anteriorly through its attachment opposite the seventh rib, near the transversalis muscle; afterwards in its free portion, so that a portion of the pleural cavity is seen; and posteriorly through its arch and crura. The pleural cavity, which is clearly evident at the posterior wall of the trunk, reaches further downwards there than it does in front; and extends on the left side to the section of the ninth, and on the right side to the seventh rib. It appears as a fine chink, which in pleurisy widens out into a considerable cavity, and may hold a large quantity of fluid (about one pound) before its presence can be demonstrated. A normal lung, however, may fill up this space in deep inspiration.

Besides the rest of the liver, stomach and spleen, in the space included by the diaphragm and the transversalis muscle, are seen the kidneys, pancreas and intestines. The section, of which the upper surface is here represented, is three and a half inches below the preceding, its inferior surface reaching to the navel.

In order to make the cavities of the intestines clear, their frozen contents were with great care broken loose before their walls were thawed by means of warm pincers; and then the cavities accurately drawn with their folds in the hardened condition. Thus the regular sharply pro-

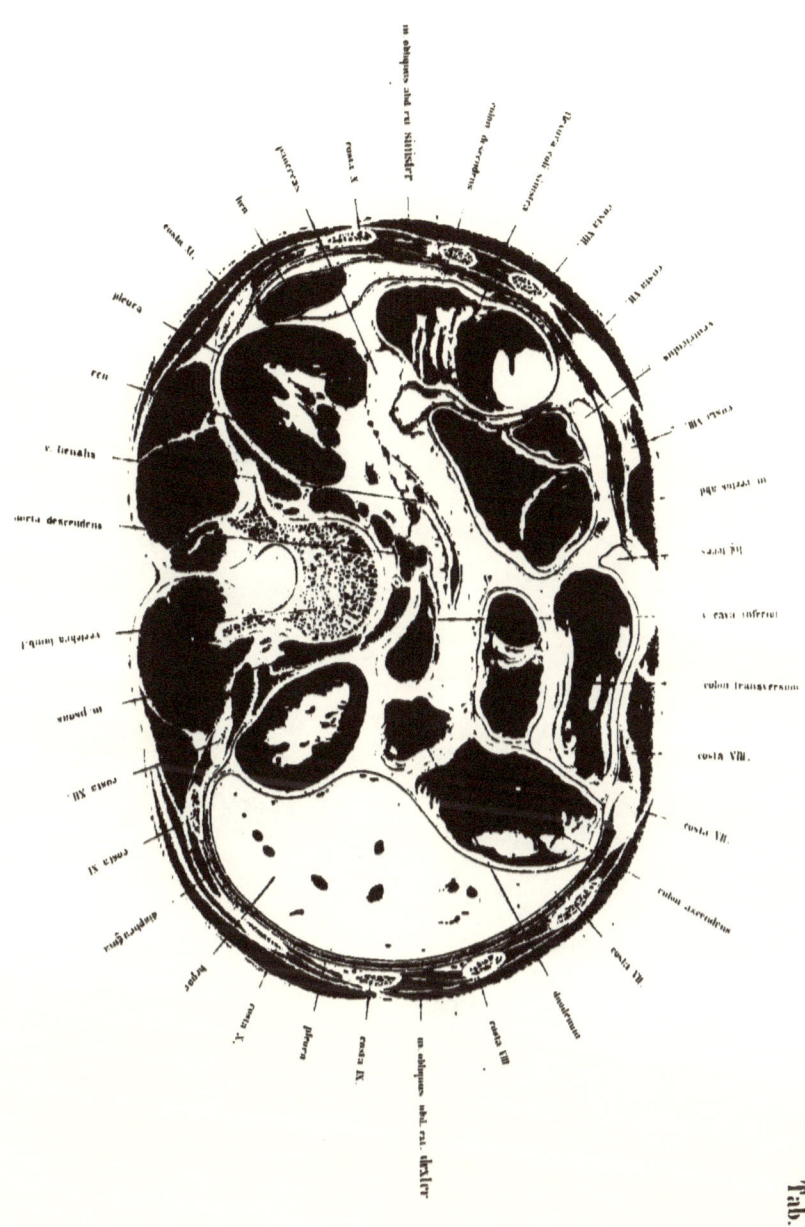

m. obliquus abd. ext. sinister

costa X.

pleura

costa XI.

pleura

ren

v. lienalis

aorta descendens

vertebra lumb. I

m. psoas

costa XII.

costa XI.

diaphragma

hepar

costa X.

pleura

costa IX.

m. obliquus abd. ext. dexter

colon descendens

Corpus vertebrae

costa VII.

costa VIII.

ventriculus

costa VII.

m. rectus abd.

hepar

v. cava inferior

colon transversum

costa VIII.

costa VII.

colon ascendens

costa VII.

duodenum

costa VIII.

Tab. XV.

PLATE XV   123

jecting folds of Kerkring, of the small intestine, and the irregular flat processes of the mucous membrane of the large intestine, are easily recognisable. On the liver, on the left side, anteriorly and internally, is the kidney, and the descending colon immediately below its left colic flexure, which is divided transversely. The contents were some green coloured fœcal matter and a little gas. Between the ascending colon and the right kidney, is the vertical portion of the duodenum, divided transversely just as it winds round the head of the pancreas. The liver fills up the remaining space externally as far as the diaphragm. Its surface has the impression of the neighbouring structures. Its convex upper surface attaches itself intimately to the line of the diaphragm; internally, on the other hand, the outline of the liver becomes irregular, owing to certain prominences in front from the impression of the colon, and behind from that of the kidney—forms still recognisable after these organs have been removed, but which, however, may disappear from the equalisation of pressure within the peritoneal investment. It is open to proof that the form of the liver is not an independent one, but varies with the pressure and volume of neighbouring organs; so that in a normal condition it must possess a softness of structure which can be compared with fat and connective tissue, and which yields to the movements and change of position of the organs in contact with it. A series of sections of frozen bodies in the region of the liver should be made, or the plates of Pirogoff (fasc. iii, 1, 2, 3, 5, 7) compared, to show that everywhere it is defined by the neighbouring organs, and entirely fills up all remaining spaces.

Only a small portion of the spleen is seen, entirely covered by peritoneum, and at this point nowhere attached thereto. Its posterior extremity reaches to the section of the eleventh rib, and corresponds also, if the preceding plate be examined, to the course and curves of the ninth, tenth, and eleventh ribs. Hence it agrees with what Luschka has recently published ('Präger Vierteljahrschrift,' Bd. 101, 1869, p. 122).

In the text to Plate XIII I have introduced three woodcuts, figs. 1, 2, 3, which explain the position of the spleen, although not originally with this view. They place its position in the upper compartment of the abdominal cavity in the cupola of the diaphragm in the different states of

distension of the stomach, and were made from preparations, of which a view
was obtained from above after raising the peritoneum by taking away the
upper half of the thorax and part of the diaphragm; and, although the
perspective view of the position of the spleen is not quite correct, it gives
the same results.   Pirogoff's plates (fasc. iii, B.), which represent plastic
preparations made by chipping the organs out of frozen bodies must be
compared with them.   It is shown in all these figures, as Luschka states,
that the spleen does not occupy the highest point of the cupola of the
diaphragm, and moreover does not lie with its hilus on the fundus of
the stomach; but that the fundus of the stomach, covered by the left lobe
of the liver, lies in the highest part of the cupola of the diaphragm, and
the spleen takes up its position laterally with it.   Correspondingly with
the greatest amount of play of the diaphragm, the position of the spleen
will not be affected; and in breathing will be less displaced than if it lay high
up in its cupola: at the same time, the influence of respiration is considerable enough to be of practical importance.   The size of the normal spleen
cannot always be felt with certainty in deep inspiration; if, however, it
be enlarged, it can be reached with the finger, on causing the individual to
take a deep breath.   The determination of the size of the spleen, by percussion, always presents certain difficulties which must not be under-rated.
Whilst on this subject I may mention that the kidney, and left colic flexure
when distended with fæces, have more than once been mistaken for tumour
of the spleen.

    A small strip only of the stomach is seen in front of the seventh costal
cartilage.   The connection between the duodenum lying between the liver,
pancreas and right kidney no longer exists.   It can be seen, however, from
the position of the duodenum that the pylorus must have lain near the
middle line of the body, and so also that the pyloric portion of the stomach
took an oblique direction from below backwards, hence the pyloric valve
could not have lain in an antero-posterior direction directly, but more
obliquely forwards (Luschka).   In Pirogoff's Atlas (iii, 2, fig. 1), is a section
which passes exactly through the pylorus and shows this relation clearly.
According to this plate the pylorus lies in the anterior half of the abdominal
cavity near the eleventh costal cartilage, immediately to the right of the

PLATE XV 125

middle line of the body. It agrees, however, exactly with Luschka's statement that the pylorus is not to be sought in the right hypochondrium, as it never reaches the right arch of the ribs; and from the present plate one can see that it must have had the same position. Hence, it follows that the horizontal portion of the duodenum does not run from left to right in a transverse direction, but more in an antero-posterior one between the ductus choledochus and the gall-bladder, close to the transverse fissure of the liver.

The duodenum is divided in its vertical descending portion at the point of flexure of the upper horizontal part. Between the vena cava and the pancreas is the ductus communis choledochus, which has passed directly over to the left side of the duodenum, in order to open into the vertical portion of the duodenum at the head of the pancreas. If we look into the duodenum we see how it curves round the head of the pancreas, becoming continuous on the left side with the inferior horizontal portion. Owing to the mobility of the stomach, without which the different stages of distension would cause great disturbance, we may expect that the pylorus and the upper portion of the duodenum would change with its condition of distension. I have proved that, whilst in the empty stomach the pylorus lies near the middle line of the body, in greater distension it is pushed half an inch further back. The upper portion of the duodenum possesses a mesentery in the hepatico-duodenal ligament, which permits and follows its changes in position. The middle or vertical portion of the duodenum is not absolutely fixed, but follows the movements of the ascending colon; and in distension is pushed to the left of the middle line, assuming its original position when the colon is emptied.

The pancreas is divided obliquely, so that a small portion of the head remains on the left side and a considerably larger portion on the right. These relations correspond with the position of the pancreas, as it does not lie exactly horizontally, but passes obliquely downwards from left to right; consequently the splenic vein, which lies below it, has had its upper surface removed in the section, and the mouths of the veins opening into it are seen.

The vein entering it directly in the middle line of the body is the superior mesenteric ; and at their junction the portal vein commences. This position is so constant that vertical sections in the middle line would expose a large portion of it (Plate I and II). The portion of the pancreas lying behind the vein is the so-called lesser pancreas.

The position of the pancreas at the level of the first lumbar vertebra, corresponds with that in Plate I and II; it is, however, so increased in breadth that it extends downwards to the next vertebra.

Behind the pancreas on the right side is the vena cava, with the left renal vein opening into it, and near it on the left side the abdominal aorta. In front of the latter passes the superior mesenteric artery, in order to gain the root of the mesentery beneath the gland.

The aorta has nearly reached the middle line, where it divides below the third lumbar vertebra into the common iliac arteries. Its distance from the anterior wall of the abdomen is nearly four and a half inches ; whilst the distance in the preceding Plate at the level of the eleventh dorsal vertebra, in the same body, is nearly five inches.

In plate, No. XVI, corresponding with the cartilage between the third and fourth lumbar vertebra, this distance is only three and a half inches ; so that it is clearly evident that the anterior curvature of the lumbar spine brings the vessel nearer the abdominal wall, rendering its compression from the front possible. ·

The section of the kidneys is such that it cuts the right above its hilus, whilst on the left side it has passed through it. The left kidney lies a little higher than the right—a relation which exists in the generality of cases. The length of the kidneys corresponds with the bodies of three and a half vertebræ, they extend from the upper border of the twelfth dorsal vertebra to the middle of the third lumbar. As they are in relation with the spleen and liver superiorly, and are bounded posteriorly by the diaphragm and pleural cavity, one would expect that they would be displaced in great pleuritic effusion by the descent of the diaphragm as are the liver and spleen. Enlargements of the liver exert a dislocating effect upon the kidney, as will be shown more exactly in the next plate.

The position of the kidneys is rather antero-posterior than transverse,

PLATE XV 127

the hilus being turned more forwards than inwards. According to Luschka, lines which pass through the hilus corresponding with the greatest breadth of these organs intersect if produced forwards, at an angle of 60° in front of the middle of the first lumbar vertebra—a statement which corresponds tolerably with the relations seen of the present plate.

PLATE XVI

THE section in this case passes through the navel dividing the soft parts just above the iliac crest, and the inter-vertebral space between the third and fourth lumbar vertebræ. The ribs are no longer seen, and the section is now below the thorax and through the middle of the abdomen. The walls of the abdominal cavity are formed, anteriorly and laterally entirely by the three oblique muscles, behind by the quadratus lumborum and the strong ligaments together with the psoas magnus of both sides. The posterior wall, where no spinous processes are visible, is very thick and strong, and formed by the mass of the dorsal muscles. The contents of the abdominal cavity are the great vessels and ureters, the ascending transverse and descending colon, and the small intestines. The contents of the intestines were carefully removed in order to allow of these viscera being accurately represented *in situ*. The section is from the same body as the preceding, and is taken about two inches lower down.

Before explaining the details here represented, I have to make some few remarks on the kidneys. They lie entirely above this section and within the region of the ribs, higher than is frequently supposed, and as many are accustomed to seek them. Hence their position may be considered as an independent one as regards the movements of the diaphragm or enlargements of the liver and spleen. I think I can prove that in both respects the relations are otherwise, and that the position of the kidneys is unchangeable.

Both kidneys extend over the bodies of three and a half vertebræ, and reach from the upper border of the twelfth dorsal downwards to the middle of the third lumbar ; and it is to be remarked that they do not lie exactly on the same level, but that the left rises somewhat higher than the right.

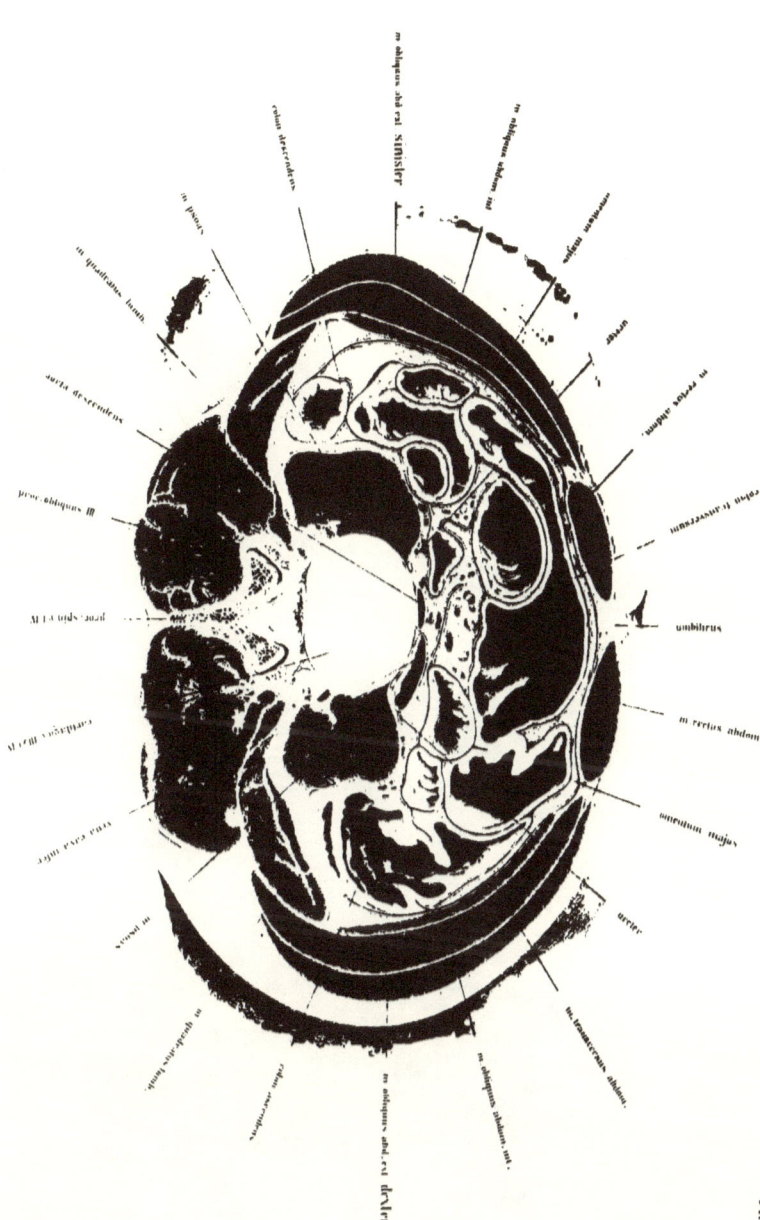

m. obliquus abd. ext. **sinister**

colon descendens

m. psoas

m. quadratus lumb.

aorta descendens

proc. obliquus III

proc. spinosus IV

M. erect. coxyges

vena cava infer.

renal

m. quadratus lumb.

colon ascendens

m. obliquus abd. ext. **dexter**

m. obliquus abdom. int.

m. transversus abdom.

ureter

omentum majus

m. rectus abdom.

umbilicus

colon transversum

m. rectus abdom.

omentum majus

m. obliquus abdom.

Tab. XVI.

PLATE XVI 129

According to Luschka (Anat., ii, 1, p. 289), they usually lie higher, viz. from the middle of the eleventh dorsal to the lower border of the second lumbar vertebra. I do not lay any stress on this, and I think that these statements may be regarded as coinciding with mine, since half a vertebra makes but little difference. The hilus lies at the level of the first lumbar vertebra, and corresponding with it is the position of the renal vessels in Plates I and II. Pirogoff gives the same (fasc. iv, tab. 4—9) ; but through the hilus in front of the first lumbar vertebra. The upper margin within which the kidneys are divided is determined by the eleventh dorsal vertebra ; the lower by the cessation of the section of the ribs, and corresponds nearly with the third lumbar vertebra.

But the relations are different if there be depression of the diaphragm, or enlargement of the liver and spleen. The kidneys are then pushed out of their position, and undergo a dislocation, which may amount to the extent of several vertebræ. In a pleuritic exudation of the right side no kidney is to be seen at the middle of the twelfth dorsal vertebra, Pirogoff (iii, 6, 3) : and in the man of fifty years, with enlargement of the liver and spleen as I have before mentioned, the hilus, as in the woodcut, fig. 1, is met with at the level of the fourth lumbar vertebra. The kidneys were also here directly pushed downwards on to the soft parts.

As regards the intestines, in Plate XVI, the inferior portion of the colon is in front; behind and on the left side the contracted descending colon; posteriorly and on the right the ascending colon more distended.

Both the ascending and descending colon lie in the angle formed by the psoas magnus and quadratus lumborum. More in the middle of the cavity of the abdomen are coils of small intestine, though not so many as one might expect. From the descending colon to the anterior border of the transverse colon is seen the cut surface of the great bag of the peritoneum passing across to the ascending colon.

It is remarkable that the intestines should show such extreme differences in calibre. According as they are empty, full, or distended with gas, they exhibit a larger or smaller cut surface. The ascending and transverse colons are large, and so also is a coil of small intestine, which has considerably compressed the end of the latter.

17

The other portions of the small intestine are only slightly distended ; and the descending colon is nearly empty.

Fig. 1.

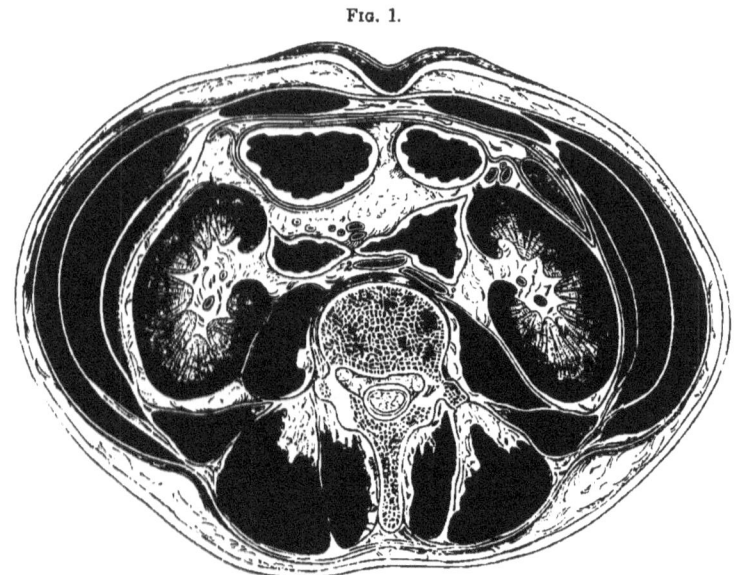

Male, æt. 50.   Dislocation of the kidneys. ½.
1. Kidney.   2. Vena cava inferior.   4. Abdominal aorta.
The vertebra shown is the fourth lumbar.

The following woodcut from Pirogoff (iii, 10, 1), which represents all the intestines fully distended, does not correspond with the natural state of things, but is the result of excessive and equally distributed artificial distension.

Pirogoff states that by inflating the intestines of a subject in all respects normal, before freezing it, he has completely distended the abdomen.

The external contour of the abdominal walls corresponded with this artificial distension of the intestines.   This contour is almost in the form of a circle, whereas mine corresponds with the normal relations, and presents

PLATE XVI   131

a flat oval.   It will be observed from the condition of the oblique muscles
how considerably the distension of the abdominal walls has compressed

Fig. 2.

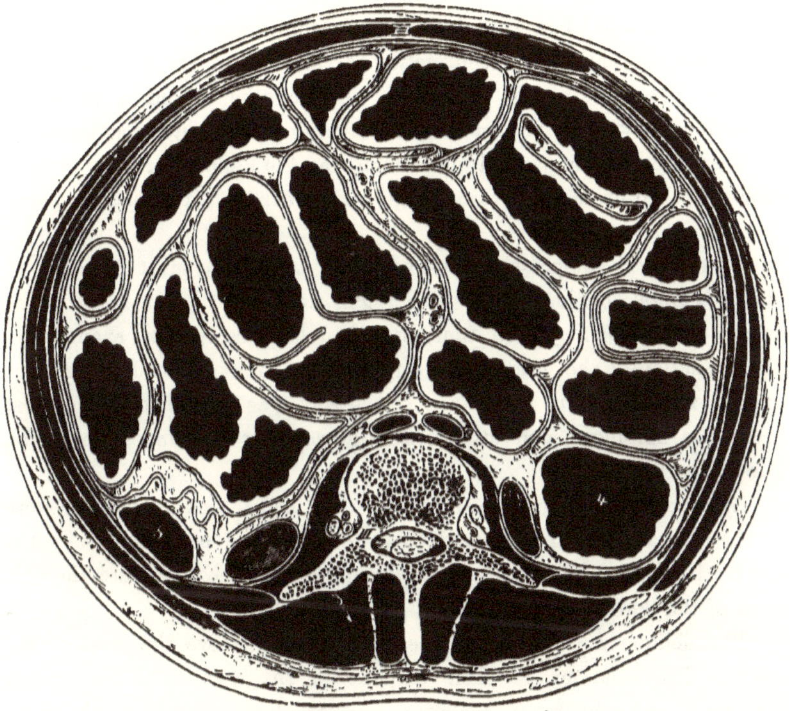

Male adult.   The intestine inflated with air and greatly distended.   Pirogoff, iii, 10, 1. ½.
1, 1. Inferior margins of the kidneys.   2. Abdominal aorta.   3. Inferior vena cava.
4. Ascending colon.   5. Descending colon.

them ; and we can estimate from their stretching and thinning the form
they must assume in pregnancy, ovarian tumours and ascites, and regulate
the depth of an incision when required.

We must notice the position of the spinal column.   As in Plate XVI
the intervertebral substance lies nearly in the middle of the circle, while in
Pirogoff's plate the position of the vertebra is far behind it.

The distance of the anterior wall of the abdomen from the spine in Plate XVI is nearly 3 inches; in fig. 1, 2·5 inches; and in Pirogoff's nearly 6 inches,—the section passing immediately below the navel.

A less distance between the spine and the abdominal walls than that shown in Plate XVI is not uncommon. This depends on the position of the diaphragm and the contraction of the lung on the one hand, and on the distension of the intestines on the other : and it is easily understood how, with normal lungs and empty intestines, the abdomen in the dead body can be pressed in so much, and the lumbar vertebræ present such a marked prominence through the abdominal walls, the distance being thus reduced to a minimum.

Therefore, in compressing the abdominal aorta, care must be taken to obtain a high position of the diaphragm, and that the intestines be as empty as possible. This compression is indispensable, for example, in disarticulation of the head of the thigh-bone. Pressure must be brought to bear immediately in the region of the navel, as the aorta divides just below the umbilicus, and still further downwards the finger would fall into the pelvis.

Lying near is the aorta in the middle line, and the cava, which is more to the side, also the ureters, and close to them and more externally the spermatic vessels. Behind and partly internal to the psoas are the sections of the lumbar nerves.

The oblique muscles are divided immediately above the crest of the ilium. The relations of their tendons to the sheaths of the rectus abdominis and quadratus lumborum are so clearly shown in the plate that we need not refer to them again. The anterior iliac spines spring forward as projections in the external contour.

It remains now to describe the position of the descending colon, and the operation for opening it, which is practicable in this region without wounding the peritoneum. This proceeding was described by Callisen, but was first performed by Amussat in 1839, and it afterwards obtained the name of Callisen-Amussat's operation for artificial anus.

This operation is preferred by most surgeons to that of opening the iliac flexure in the left inguinal region (Littre), as the descending colon has

PLATE XVI   133

a fixed position, and, being incompletely invested by peritoneum, an incision can be made into it without wounding this membrane. It is usually stated that the descending colon lies along the outer border of the quadratus lumborum ; and, in conformity with this, an incision is to be made vertically along the outer border of this muscle. This is not always correct. At the lower border of the kidney the colon lies further outwards than it does in the neighbourhood of the ilium ; and, the quadratus lumborum being narrower above than below, the rule is true as far as regards the level of the third lumbar vertebra, but not so for the deeper regions. At the level of the symphysis between the third and fourth vertebræ, and at the fourth below the kidney—and therefore exactly in the field of operation—the quadratus lumborum covers in the colon posteriorly, and must be cut in order to reach it. It is only when much distended, a condition which is not so constant as one would expect in operations, that the intestine increases in breadth forwards and inwards, or overlaps the outer border of this muscle (Pirogoff, iii, B., tab. 14). Consequently the incision, which is to be directed along the border of the great extensors of the trunk from the ilium to the twelfth rib, would divide the strong tendons of the transversalis until the quadratus is exposed, and subsequently the fibres of this muscle, when the extra-peritoneal fat and cellular tissue would be met with.

When the surgeon has carefully arrived at the cellulo-fatty tissue through the fascia beneath the quadratus lumborum, making the incision of an equal length with the primary one, so as to avoid a funnel-shaped wound, the main point is to fix the colon at its free surface and to open it. In doing so he must avoid the kidney, which from its deep position (cf., fig. 1) can easily obstruct the field of operation, and which must therefore be carefully pushed on one side. From the impossibility of recognising the peritoneum from its posterior aspect, success can only be safely calculated on by measuring the distance of the point of reflection of the peritoneum, and how far from the colon this position is constant.

In the first place, as regards the descending colon, which I here particularly refer to, after measurements on frozen bodies of full grown men, I find that this distance, in a straight line (therefore not corresponding with the curvature of the wall of the intestine), is from four fifths of an inch to one inch, supposing the intestine empty and contracted (at a level between the third and fourth lumbar vertebræ); further, that the free side of the intestine, as in Plate XVI, does not look posteriorly but somewhat inwards, exactly towards the angle which the psoas and quadratus lumborum make with each other. If, on the other hand, the small intestines are much distended, the peritoneum between the psoas and colon would be pushed further downwards; and the colon, by means of the traction of the parietal portion of the peritoneum, would be rotated on its axis, so that its free surface would be directed more outwards.

Should the colon itself be distended, its surface free of peritoneum becomes considerably larger, and may assume a breadth of from 2 to 2·5 inches. Tympanitis of the small intestine appears to have a rotatory influence on the distended colon; and on comparing Pirogoff's plates it is shown with its free surface turned somewhat outwards (cf. Pirogoff, iii, B, tab. xiv).

In the performance of the operation of colotomy a distended abdomen will probably often be met with. I therefore do not consider these remarks superfluous, and I hope that they may contribute to make the avoidance of the peritoneum more certain than heretofore where it was so much left to chance; and, as a third part of the cases show wound of this membrane, the value of Amussat's method appears problematical.

Tab. XVII.

PLATE XVII

In order to bring the pelvic organs into view, a section was made of the trunk just over the symphysis pubis, and through the lower portion of the sacrum. The section passed through the inguinal region, the outer mass of the muscles of the thigh, the head of the thigh bone near its middle, the pelvis, bladder, rectum, and some coils of intestine lying in Douglas's pouch. The ischia were divided in the tuberosity, so that the section nearly followed the sacro-spinous ligament.

The plate moreover shows, enclosed in the bony pelvis, the obturator internus and levator ani muscles, and laterally the ilio-femoral articulation with its muscles and vessels.

We notice at first that the central portion is bounded by the pubis, ischium, levator ani, sacro-spinous ligament, and the last portion of the sacrum.

The bladder, which contained about four ounces of frozen urine, appeared so contracted on its contents that its form was not affected by the pressure of the neighbouring organs, as is so frequently observed in Pirogoff's plates, whence the upper wall appears considerably fallen in. The body was perfectly fresh when brought in for preparation, and as no decomposition had set in gas had not formed, so that the forms of the cavities were not changed. The contents of the bladder were removed before the drawing was made. The internal orifice of the urethra is clearly seen in the middle of a fringe formed by folds of mucous membrane. More in front is the anterior wall of the bladder, flattened by the pressure of the symphysis. The thickness of the wall of the bladder, considering the amount of distension, is considerable. The thickness of the posterior wall, however, is due to its having been cut obliquely. In order to compare the position and form of the bladder with the section given in Plate I, the mass of ice was carefully removed and represented in profile. This could readily be accomplished, as only a small part of the upper wall of the bladder and its contents was removed with the upper portion of the section. By comparing this with the sagittal section of Plate I a close agreement in form was observed ; though they differed in the fact of the internal orifice of the urethra in Plate I being somewhat higher than in this. In both cases, however, the form and position of the bladder of a young powerful man is defined, as can be verified by injecting tallow either by the urethra or the ureters. It is, at least, certain that the spherical form represented by Kohlrausch is not a natural condition, as he omits to notice a neck to the bladder, which is a funnel-shaped contraction of this viscus towards the urethra. For a wider distension of the bladder there is, as the plate shows, ample room. The cellulo-fatty tissue on both sides of it gives way readily, and the coils of small intestine are easily lifted up and pushed on one side by the swelling bladder. The rectum will be more flattened, and room is afforded by the emptying of great venous plexuses, until at last the bladder alone almost fills the pelvic cavity. With these changes in volume of the bladder, the relations of its peritoneal coat alter. Even in the slight degree of distension shown in the present instance, only the upper wall and a small portion of the posterior were covered by peritoneum, so that there was a passage above the symphysis although but a slight one ; and it is evident that this sub-peritoneal passage must acquire breadth with the increased distension and elevation of the bladder. Behind the bladder is a flat section of the peritoneal sac containing a portion of small intestine divided behind the fold of Douglas, and behind this again is a cul-de-sac of peritoneum, the so-called pouch of Douglas. This is directed in an oblique direction forwards and downwards, and is about three fourths of an inch deep. It held about three fifths of an ounce of frozen water.

The vesiculæ seminales, which lie immediately below the section, were exposed by taking away some cellular tissue ; towards the middle line the vasa deferentia take a sharp curve forwards and upwards, and are divided in the section ; their small calibre and thick walls are well seen. Anteriorly and somewhat externally, are the ureters in section. The

PLATE XVII   137

rectum, which contained a little fæces, was divided shortly before its final curve. The anal extremity was fully 3-8 inches distant from it. If the peritoneum leaves the anterior surface of the rectum entirely free, and, under the form of Douglas's pouch, descends here externally and internally rather more than half an inch ; this pouch is about three inches from the anal aperture, and it follows that at this level an operation on the rectum might be undertaken, without fear of wounding the peritoneum. These relations correspond with those of Plate I.

The question arises with regard to the rectum, as in the case of the bladder, what changes of form it assumes with its varying degrees of distension ; that it is capable of very great changes in volume, both experience and experiments by means of injection show us.   The requisite space is provided for in the same way as for the bladder ; the cellular tissue and fat are pushed aside, Douglas's pouch and the intestines are lifted up, and in fuller distension of the rectum the bladder is raised somewhat upwards and forwards.   The following woodcut from Pirogoff's atlas is instructive on this point.

There is little here that needs explanation ; the great similarity in form with my plates will facilitate the description.   More than half the cavity of the pelvis is occupied by the distended rectum, which is cut through about two inches above the anus, and is considerably distended with air.   The semi-lunar fold has not been obliterated by this distension, but springs up sickle-shaped into the cavity.   The contour of the pelvic cavity is worthy of notice.   The section passes as above mentioned through the spines of the ischia, and partially through the sacro-spinous ligaments, and between the ischiatic notches.   Corresponding with it, a process of bone springs from the body of the ischium on both sides, tolerably far backwards, and terminates in the whole length of the sacro-spinous ligament as far as the sacrum.   On the right side this band is only to be followed for a certain distance from the sacrum, and does not reach to the apex of the ischium as on the left ; while the sacro-spinous ligament has a horizontal direction, the tuberoso-sacral ligament has a more vertical one, and a small portion only of the latter is seen.   It is shown near the sacrum at the edge of the gluteus, where it deviates from the other ligamentous band and lies deep.

18

Between these fasciculi, on the left side, lie the internal pudic vessels and nerves; on the right they are further off, and are to be looked for near the spine of the ischium. Internally on either side from the sacro-spinous ligament is a dark band, partly prolonged to the spine of the ischium and partly associated with the fascia of the obturator internus; this is the superior portion of the levator ani. This muscle closes-in the cavity of the pelvis like a muscular funnel, and consequently may not be inaptly com-

Fig. 1.

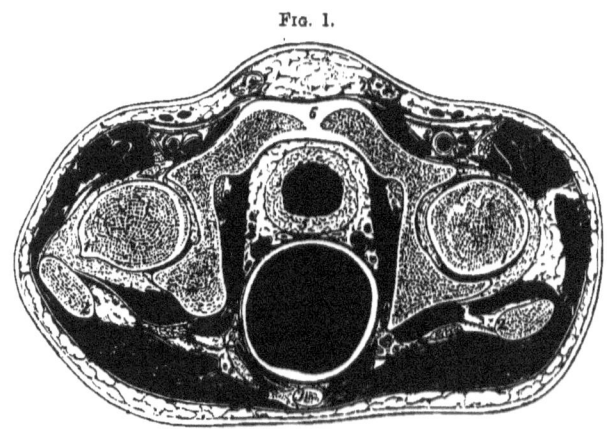

Transverse section of the pelvis of a boy, æt. 15. Pirogoff, fasc. iii, tab. xvi, fig. 1.
1, 1. Head of femur.   2, 2. Great trochanter.   3. Tip of coccyx.   4. Rectum distended with air.
5. Bladder.   6. Upper border of symphysis pubis.   7, 7. Spermatic cord.   8, 8. Femoral
vessels.   9, 9. Obturator internus.   10, 10. Gluteus maximus.

pared to the diaphragm. All sections which divide the bladder further downwards must therefore fall within the region of this muscle, and expose it as a muscular ring limiting the pelvic organs. Such a section is shown in the following figure.

Fig. 2 represents a section that I made on the pelvis of an old man. It passes through the symphysis; on the left side through the lesser sacro-sciatic foramen; on the right somewhat lower, through the tuberosity of the ischium; and posteriorly through the tip of the coccyx. The levator ani is seen bounding the pelvic cavity, which contains behind the rectum a coil

PLATE XVII   139

of small intestine, the vesiculæ seminales, and the neck of the bladder and urethra.

Since this section is taken considerably deeper, the left gemellus inferior is seen running in direct relationship with the obturator internus muscle; notwithstanding this, Douglas's pouch with its peritoneum is present. It thus appears that the position of the peritoneal sac is deeper in the present one than on the young subject shown in fig. 1.

If we consider, moreover, that in new-born children the position of

Fig. 2.

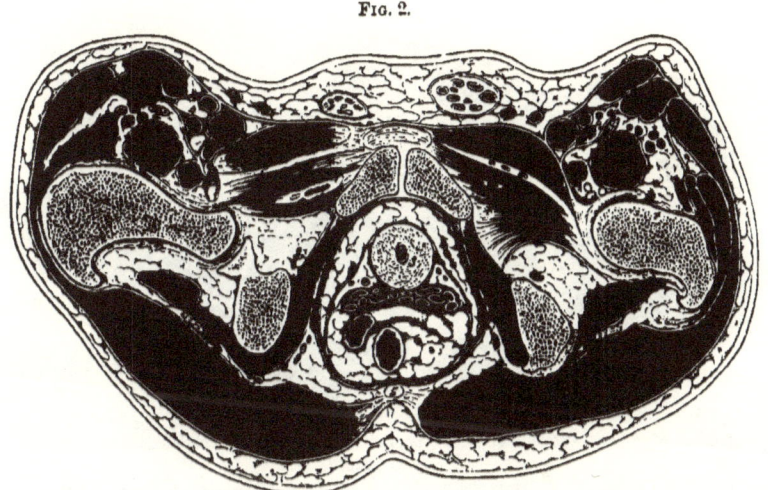

Transverse section of the pelvic cavity.

1, 1. Head of femur. 2. Rectum. 3. Bladder. 4, 4. Femoral vessels.
5. Apex of coccyx. 6. Gluteus maximus.

the peritoneum relatively within the pelvis is particularly high this relation must be described as natural and corresponding with advanced age; and hence one must be particularly careful, in operations on the rectum in old people, not to wound the peritoneum, which extends lower down than in younger individuals.

Fig. 3 is reduced from Pirogoff's atlas. It is stated in the text (fasc. iii, p. 59) that it was from the body of a full-grown man, whose bladder and rectum were full. The section passed through the lowest portion of the symphysis, about seven lines below its upper border, through the foramen ovale, the tuber ischii, near the lesser sciatic notch and the insertion of the sacro-sciatic ligament, and included the coccyx posteriorly. The lower half of the section is represented, so that it is viewed from above downwards.

In such a section, a clear view of the levator ani cannot be given, at

FIG. 3.

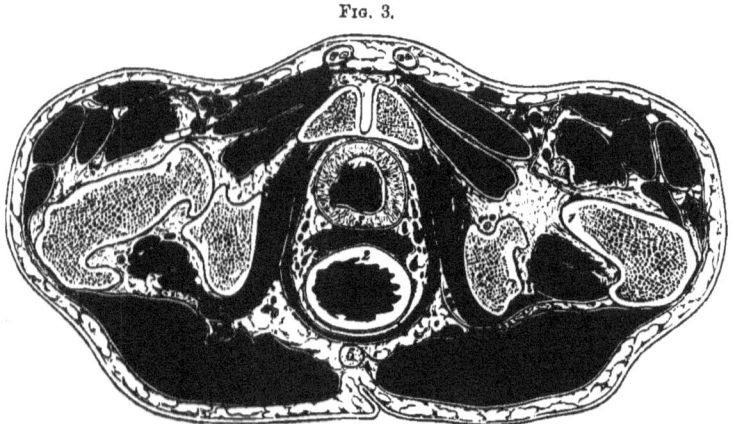

Section of the lower portion of the pelvis of a full-grown man, with distended rectum.
Pirogoff, iii, 16, 3.
1, 1. Head of femur.   2. Rectum.   3. Bladder.   4. Femoral vessels.   5. Tip of coccyx.
6, 6. Gluteus maximus.

least not with respect to its physiology, as only a small portion of its fibres would be divided. It will be clearly seen at the same time that when the rectum is full Douglas's pouch and the lower coils of the smaller intestine are lifted up. Between the bladder and rectum lie the sections of the vesiculæ seminales. Outside the pelvic cavity, are the thigh bones divided through the neck, with the ligaments and their corresponding vessels and muscles. As the head of the thigh bone presents a spherical form only internally and above, so each transverse

PLATE XVII 141

section which passes through it near its middle includes a portion of its neck, and therefore produces externally a cut surface very far removed from a circle. The inner contour only would present a portion of a circle, namely at the point of insertion of the ligamentum teres. The component portion of the joint is better seen higher up.

Fig. 4 is taken from a series of sections on the body of an old man,

FIG. 4.

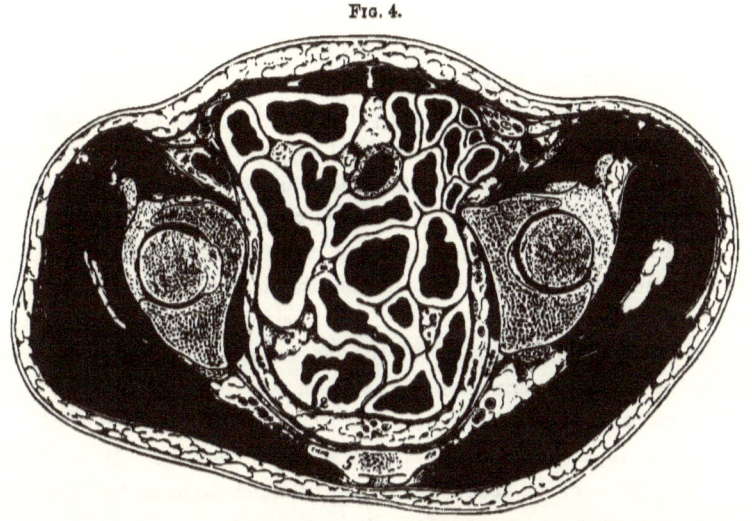

Section of the pelvis of an old man at the level of the great sacro-sciatic ligament.
1, 1. Head of thigh bone. 2. Rectum. 3. Apex of bladder. 4, 4. Femoral vessels.
5. Lower end of sacrum. 6, 6. Gluteus maximus.

somewhat higher than Plate XVII, and is therefore useful in comparison with it, since it traverses the entire length of the sacro-sciatic ligament.

The acetabula are divided nearly through their centre. Nothing more of the symphysis pubis is to be seen, as in consequence of the greater inclination of the pelvis it lies considerably deeper. The relation of the vasa deferentia to the femoral vessels is well shown. Corresponding with the deeper position of the viscera in old persons already mentioned, a quantity of coils of small intestine are here shown, whereas on Plate XVII there is merely a small flat section of the ileum.

PLATE XVIII

I⊤ appears to me desirable to introduce here a frontal section of the pelvis, and one that will show the relations of the hip-joint to the best possible advantage. After many investigations, I became convinced that for this purpose a definite position of the bones is necessary, as when the subject lies on the back they are rolled outwards, and the head, neck, and shaft do not lie in the same plane. It is only when the thigh is rolled considerably inwards, so that the inner borders of both feet touch throughout their entire length, that they do so; I made the section, therefore, with the feet tied together.

The section passed through the pelvis and hip-joint in such a manner as to render the two sides as symmetrical as possible. The upper portion of the shaft of the right femur is not divided quite in its axis, and only a portion of the great trochanter is clear, while the lesser trochanter is covered with muscles. The head and neck are fairly divided. The section passed through the middle of the acetabulum; through the whole length of the ligamentum teres of both sides, the obturator foramen and the ilium. The promontory of the sacrum and the tuberosities of the ischium lie in the posterior half of the body.

The preparation is viewed from the front, and thus the right side of the body is to the left of the picture and the converse. It represents the lower portion of the abdominal cavity, bounded above by the three flat abdominal muscles, and more externally by the iliaco-psoæ, in which are the anterior crural nerves. Within these muscular walls are the intestines, extending as far down as the bladder, the anterior portion of the cavity of which is opened. The sections of the small intestine, which above is

Tab.XVIII.

PLATE XVIII   143

jejunum and below ileum, as can be readily recognised from the nature of
their mucous coats, indicate that in many instances they have been met
with in their long axis. There are singularly few instances in which this
has happened in the preceding sections, and it therefore follows that the
coils of intestine have a parallel direction with the long axis of the body.

Of the individual portions of the intestine, the section of the vermiform
process is seen at the upper border of the right psoas ; and on the left of the
iliac vein the transverse section of the rectum. The latter was especially
studied in relation to its course. It ascended behind Douglas's pouch, in
the left half of the body near the middle line ; curved sharply forwards over
the left psoas muscle, so that it fell in the plane of the section ; and then
passed somewhat forwards towards the right half of the body as an
arc of a large curve, ultimately becoming continuous with the descending colon. It shows, moreover, a deviation from the usual course,
at the lower portion, as figured by Pirogoff (fasc. iii B, tab. xv, fig. 1), but
does not completely correspond with the relations shown in Plates I and II;
and one can easily convince oneself by injecting with tallow that, in individual
cases, and those not very rare, the S-curve of the rectum is not sharply
marked in a frontal direction with regard to the sacrum,—variations which
are owing to the inconstant length of the meso-rectum. Should this be
strong and reach far back, the position of the rectum is freer, and more
dependent on the condition of the neighbouring organs. Shortness and
tenseness of this meso-rectum, on the other hand, contribute to a firm and
constant position of the intestine.

The effect produced by the distension and by the firmness of the walls
of the rectum must be taken into consideration. Great distension from
fæces, and flaccidity of its walls especially, permit of considerable stretching of the original curves. It can be proved by investigation and clinical
observations, that the surgeon can straighten the curved rectum by means
of instruments, and introduce them as far as the iliac flexure. Foreign
bodies introduced from the anus, and firmly impacted, can be seized with
forceps and withdrawn.

The bladder contained a little urine, and was firmly contracted : it is
separated from the section of the levator ani by a little fat ; on both sides

of the levator ani lie the sections of the obturator internus, bounded below
by the obturator membrane, and laterally by the pelvic bones. If the space
between the intestines and the pelvis be followed upwards on both sides
from the bladder, beneath the peritoneum, we meet with two whitish oval
sections, which represent the lateral ligaments of the bladder. They lie
thus far removed from the bladder, because it was small and contracted ; a
distended bladder would carry them upon its upper surface, and at the
same time occupy the entire space of the inferior aperture of the pelvis, as
several of Pirogoff's plates show. Farther outwards, and in the same space,
between the peritoneum and the pelvis, is the vas deferens, and above it
the obturator vein and nerve and a small artery. The main trunk of the
artery passes through the obturator foramen.

    Finally, we arrive at the external iliac artery and vein ; both vessels
lie on the inner wall of the psoas, as the preceding sections show, not side
by side, but behind each other ; hence the artery lies over the vein, and not to
its inner side as appears by this frontal section.

    The relations of the hip-joint, which have been already briefly alluded to,
afford many points for examination. It has been already mentioned that
the section has traversed the entire length of the ligamentum teres of both
sides. It is evident that this ligament limits extreme adduction, and by
simultaneous stretching, assists in maintaining the firm position of the pelvis
and trunk. As the section passed through the acetabular notch the course
of the articular artery is exposed.

    The articular cartilage, ligamentous apparatus, and the extent of the
cavity of the joint are well seen in the plate.

    The architecture of the upper portion of the thigh bone is well worthy
of study, as much so for its general disposition as for its structure. Meyer
has the merit of having first called attention to the arrangement of the
cancellous tissue, especially in the neck of the bone, which essentially
increases its weight-bearing power. The individual laminæ and interlace-
ments of bone arrange themselves in rows, which are detached from the
borders of the compact tissue, and cross each other in the middle line. In
the section of the left thigh bone especially these indications of its structure
are shown.

PLATE XVIII   145

The articular cavities themselves appear merely as chinks. Their extent downwards explains to what limit intracapsular fracture of the neck of the thigh bone may reach, and where the region of extra-capsular fracture commences. Since intra-capsular fractures isolate the upper fragment, and leave it connected by the ligamentum teres and the acetabular vessels, it is evident that, apart from the difficulty of accurate adaptation and retention of the parts, union is of very rare occurrence, on account of conditions unfavourable for its nutrition.

An increase of effusion into the joint, as may happen in inflammation, will not separate the surfaces of the acetabulum and head of the thigh bone. The powerful ilio-femoral ligament, in consequence of its torsion in complete extension, presses the joint-surfaces firmly against each other. On the other hand, in flexing the joint, a corresponding separation of the two surfaces will occur from increased effusion within it; and, as investigations show, this may be somewhat considerable. If fluid be injected through the acetabulum into the joint-cavity, after the example of Bonnet, the articulation takes successively the positions which afford the greatest amount of space; but which ultimately place the ilio-femoral ligament in the condition of greatest relaxation. The femur is raised and somewhat rolled outwards. If the joint be frozen, sections can be made of it, and the relations of the articular surfaces to each other rendered clear. The accompanying woodcut represents such a preparation, made from the body of a normal young female.

In order to render the femur more easily movable, the upper layer of muscles was removed and the bone itself sawn through the middle.

On injecting the joint with tallow, and applying as great a pressure as possible, the femur was raised and rolled outwards. In this position it was frozen and sawn as shown in the woodcut; the section passing not quite through the middle of the head, but slightly in front, and including the trochanter minor in its course. The mass of tallow, which is here represented by the dark shading, was about one fifth of an inch thick, and a little farther down in the articulation somewhat thicker; and surrounded the head of the bone like a cap, extending outwards to the attachment of the synovial membrane, which was driven forwards in the form of a bladder on its posterior wall. We should expect to find, in diseases of the hip-joint which exhibit similar positions of the articulation, an actual lengthening of the thigh, supposing that a like quantity of fluid exists in the joint cavity. To prove this by measurement is impracticable. Were it possible to measure it accurately to a quarter of an inch, which from the simultaneous displacement of the pelvis can hardly be expected, the flexion of the thigh, associated with this condition, renders such measurement impracticable.

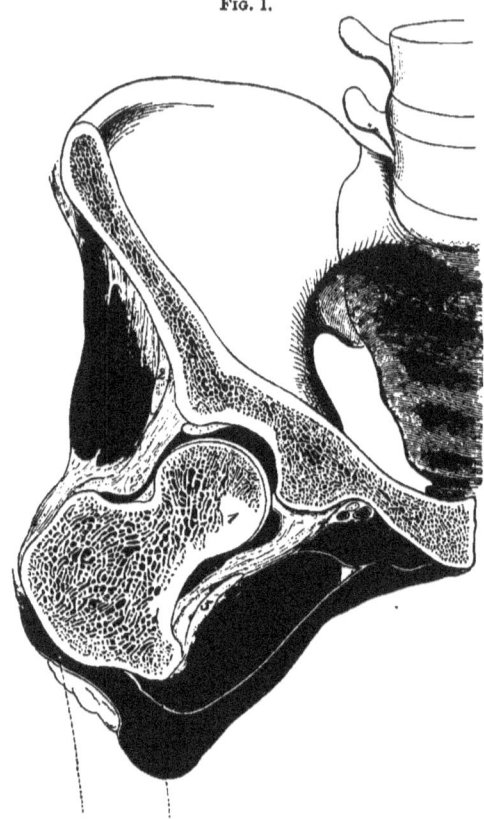

Fig. 1.

Frontal section of the hip-joint injected with tallow and frozen. ⅓.

1. Head of femur. 2. Tendon of rectus. 3. Obturator externus. 4. Pectineus. 5. Tendon of ilio-psoas. 6. Gluteus minimus.

The relations of the corpora cavernosa and urethra next demand attention. It will be seen that the section passes in front of the prostate, dividing the corpora cavernosa penis near their origins, and the urethra at the bulb.

The corpus cavernosum, arteries, and muscles of the corpus cavernosum are well shown. Upon it is expanded a portion of the deep perineal muscle with a number of large veins.

PLATE XVIII

147

FIG. 2.

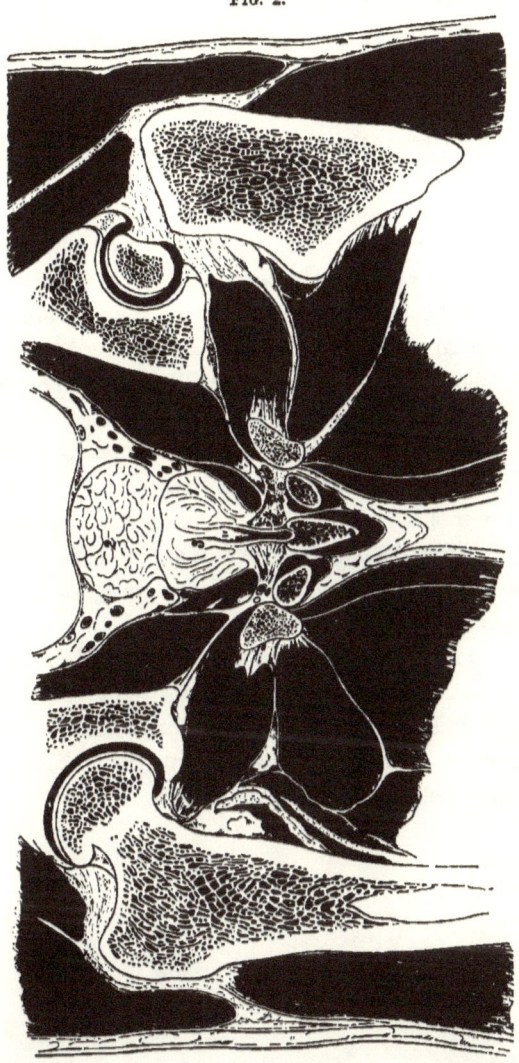

Frontal section of the male pelvis through the membranous portion of the urethra. ½.

1. Prostate. 2. Wall of bladder. 3. Caput gallinaginis. 4. Deep transversus perinei muscle.
5. Bulb. 6. Ascending ramus of the ischium. 7. Obturator membrane. 8. Obturator externus.
9. Obturator internus. 10. Adductor magnus.

As it appeared to me desirable to have a section showing these structures rather farther back, I made one on the body of a normal well-built man, at such a depth as to pass through the prostatic portion of the urethra. The preceding woodcut represents the plate on a smaller scale. The head of the left femur is seen only as a small segment, and not in connection with the rest of the bone.

The body of the ischium shows a large surface in section, corresponding with its more extensive development behind the acetabulum. The obturator membrane, ascending ramus of ischium, and the obturator externus and internus, still show some resemblance to the corresponding portions on Plate XVIII, and so also do the corpora cavernosa. We have, moreover, in the section, in place of the apex of the bladder, its posterior wall, and the posterior half of the prostate, with the caput gallinaginis.

The membranous portion of the urethra and the prostate are opened. On both sides of it are the deep transverse perineal muscles, the fibres of which are expanded towards the middle line. Above is seen the anterior mass of the levator ani. Around it is a layer of fascia, the upper portion of which is continuous with the pelvic and the lower with the perineal fasciæ. Both fasciæ meet at the inner border of the levator ani muscle, and help to support the prostate. The upper lamina of the perineal fascia and the lower surface of the transversus perinei pass forwards.

The plate, which must not be regarded as diagrammatic, agrees in all its essential particulars tolerably accurately with Henle ('Eingeweidelehre,' p. 504, fig. 392), which should be compared with it.

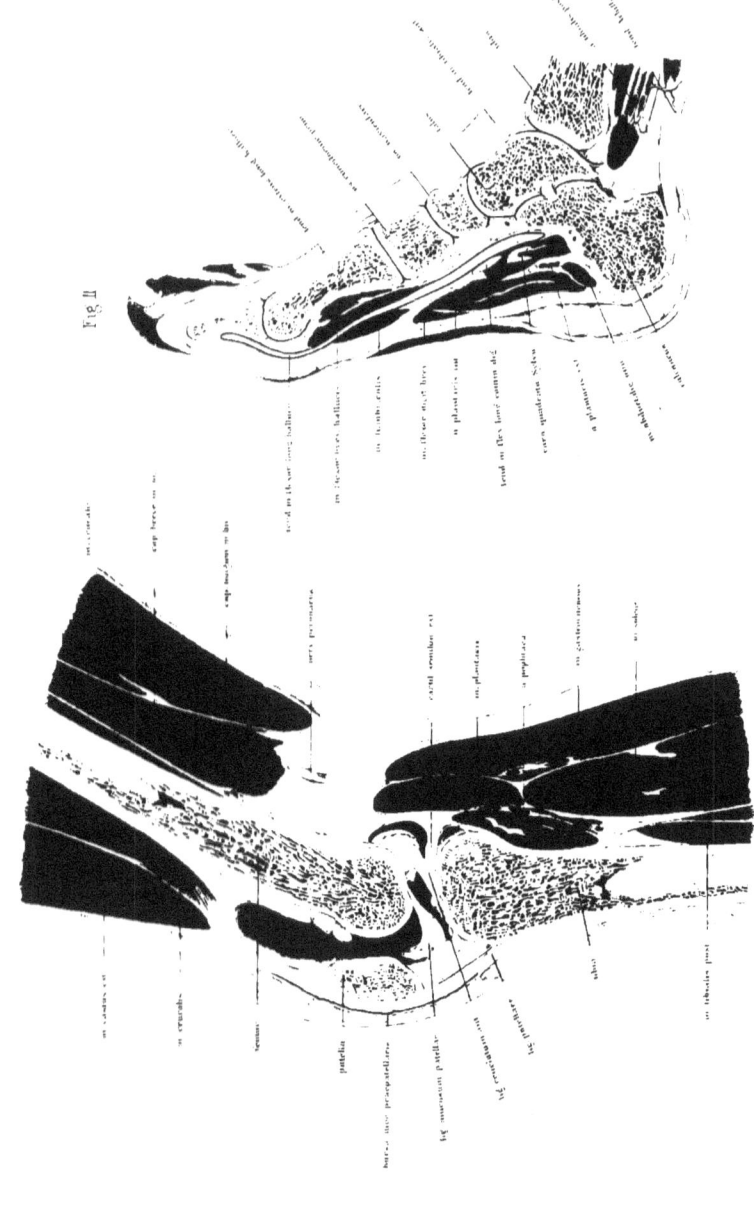

Fig II

PLATE XIX

In order to demonstrate the shape of the cavity of the knee-joint and the extent of its capsule correctly, I injected water into the articulation with a Pravaz's needle under great pressure, and, having slightly flexed the joint, froze it. The limb was taken from a normal body (young female). The section passed tolerably nearly through the middle, and divided the extremity into two nearly equal halves, of which the right one was used for the plate, after the removal of the frozen water.

All the joints, not the hip and shoulder only, are subject to atmospheric pressure; and, on account of the small quantity of synovia which they contain, can retain their normal position and not show free cavities, as one finds on opening a joint in a soft preparation. Accordingly the synovial cavity appears in the section of a normal joint as a narrow crevice; which in the following section of a normal uninjected knee-joint is represented by a single black line.

If this joint be compared with the injected specimen, as represented in Plate XIX, one can understand the meaning of the black line which indicates the joint cavity. Further, the position of the patella is seen in normal and abnormal joints. Whilst in the normal condition of the joint the patella touches the femur with a small portion of its cartilaginous surface like a tangent, in the case of the distended synovial membrane it is completely lifted off it. The patella floats, supported by the fluid as a board on water, and must therefore yield under pressure of the finger until it reaches the femur, which lies behind it.

The capacity of the joint-cavity is well shown, whilst in the woodcut the synovial membrane of the extensor muscles appears as separated from it, since the wide aperture of communication which unites it with the bulging out of the capsule is not opened by the section; and on the injected joint represented in Plate XIX no such separation is to be seen. The fluid injected has penetrated into all the portions and hollows of the joint, and has raised up the posterior wall of the capsule, so that the posterior portion of the condyle of the femur is brought into view.

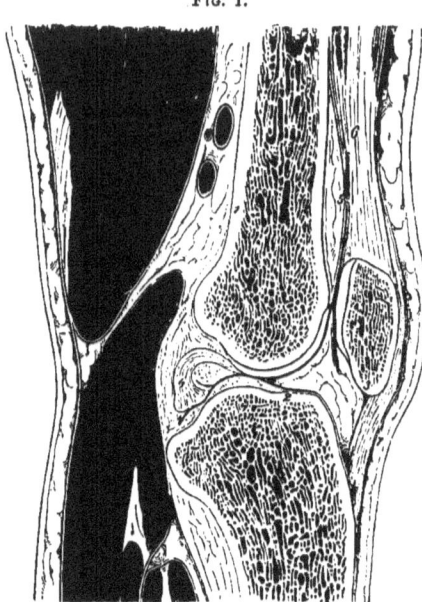

'Fig. 1.

The ligamentum mucosum of the patella and the anterior crucial ligament lie in the plane of section.

It is well known that Bonnet was the first to apply the method of injection to the investigation of joints, and to prove thereby what position of the joint corresponded with the greatest distension of the synovial cavity. It appeared in all joints that it was the position of flexion that allowed of the greatest amount of fluid entering the articular cavity; and that, with strong pressure of injection, all joints, no matter what position they may have

Longitudinal section of the frozen knee-joint of a full-grown man. ⅓.
1. Femur. 2. Tibia. 3. Patella. 4. Posterior crucial ligament divided. 5. Bursa mucosa. 6. Quadriceps extensor. 7. Ligamentum patellæ. 8. Semi-membranosus. 9. Gastrocnemius.

had beforehand, acquire the position of flexion and maintain it so long as the pressure is continued. It is natural to suppose also that in diseases of the joint, associated with effusion into the synovial cavity, the position of flexion which the patients involuntarily affect is brought about by the direct pressure of the fluid.

But against such a supposition the following points may be adduced, as can be well explained after consideration of this plate.

The capacity of the joint-cavity also depends on the possibility of the

PLATE XIX   151

separation of the patella which is developed in the extensor tendon from the surfaces of the condyles. This is, however, the case when the extensor tendon is relaxed, as in extension, or in only slight flexion of the joint; in greater flexion the patella must be pressed against the condyles, by the tension of the quadriceps, thus causing a diminution of the capsular cavity. It will therefore be expected that in consequence of the extension of the synovial space upwards beneath the extensor tendon, a considerable quantity of fluid may be injected, and that a greater degree of flexion must directly diminish the amount. I therefore considered it necessary to undertake a repetition of Bonnet's researches with the greatest possible care, and that on entire bodies. The method I used was the following :

The subject was fresh and normal, and, after violently breaking down the rigor mortis of the lower extremities, was laid on its back on a horizontal table. The thigh hung down over the free edge, and during the investigation was fixed by means of a support under the heel by an assistant in the necessary position. A screw was driven into the upper third of the tibia, to the free extremity of which a flat piece of wood was fastened; which served to fix a dial plate, provided with a graduated semicircle; and it was so arranged that a plumbline fastened to the centre of the circle stood at zero in complete extension of the bone, and the amount of flexion could be immediately read off. No regard was taken of the rotation of the thigh during flexion. In order to prevent diffusion through the capsule, the fluid used for injection was a solution of common salt, contained in a graduated tube about sixty inches in length, to the inferior end of which was fastened a short piece of tubing of india rubber, carrying a strong Pravaz's needle. The tube was fixed in an oblique position by means of a movable support, so that the vertical line, indicating the difference in height of the point of introduction of the needle and of the level of the fluid, always remained the same; by which means the pressure indicated by the constant height of the support was maintained. The apparatus thus formed a right-angled triangle whose hypothenuse was represented by the obliquely directed tube, the perpendicular by a portion of the support, and the base by a horizontal line running parallel to the table and extending from the point of introduction of the needle to the support. The point of introduction of the needle being as near as possible in the axis of rotation, it remained almost unaltered in flexion of the knee-joint; consequently it was possible from the changing level of the water in the tube, to read off the diminution or the increase of the fluid in cubic centimètres. Of course, the support had to be constantly placed under the meniscus of the fluid, whilst the zero point of the tube was kept in an unaltered position relatively with the point of introduction of the needle. Thus, whilst the side of the triangle indicating the pressure was constant, the length of the hypothenuse and that of the other side varied, becoming larger on diminution of the volume of the synovial space, and smaller in the contrary condition.

By means of this method of investigation it was possible to determine the following points which Bonnet's proceeding could not afford. We could immediately ascertain the dependence of the capacity of the synovial cavity on the angle at which the bone was placed, since the pressure of the fluid in the walls of the capsule always remained one and the same, and these in the intact condition of the body and extremity presented their original relations to skin, fat, muscle, &c. Thus the grade of flexion, in which the synovial cavity reached the maximum of its capacity (described by Bonnet as the mid-position between flexion and extension), could be accurately recorded. Finally, the volume of the synovial space during the different positions of the bone could be measured by cubic centimètres. The following figures, which indicate each angle of flexion, will be easily understood after the preceding description. 0° corresponds with complete extension, 10° would indicate that the thigh made an angle of 170° with the leg, &c.

The figures referring to the volume give the quantity of fluid in the capsule in each case, in cubic centimètres; those referring to pressure, in centimètres.

Experiment 1.—Body of a man, æt. 50; tolerably recent. Muscular development and nourishment good. The rigor mortis of the limb forcibly broken down. Pressure 19 centimètres.

PLATE XIX   153

Angle   . 0° 10° 20° 30° 40° 50° 60° 70° 80° 90° 100°

Volume   . 312 328 332 331 330 326 316 303 283 265 255 c.c.

Experiment 2.—Body quite recent. No rigor mortis. Pressure 23 centimètres.

Angle   . 0° 10° 20° 30° 40° 50° 60° 70° 80° 90° 100° 110°

Volume   . 114 128 137 141 141 140 135 125 112 99 76 75 c.c.

Experiment 3.—The opposite knee of the same body. Pressure 34 centimètres.

Angle   . 0° 10° 20° 30° 40° 50° 60° 70° 80° 90° 100° 110°

Volume   . 83 95 104 111 110 109 107 93 91 83 66 54 c.c.

Experiment 4.—Body of a man, æt. 50; eight days dead, poorly nourished. Rigor mortis forcibly broken down. Pressure 14 centimètres.

Angle   :   0°   10°   20°   30° 40° 50° 60° 70° 80° 90°

Volume   . 143½ 149½ 154½ 146½ 139 136 118 102 88 78 c.c.

Experiment 5.—Body of a muscular man, æt. 36´; rigor mortis broken down.

Angle   . 0° 10° 20° 30° 40° 50° 60° 70° 80° 90° 100°

Volume   . 79 90 98 104 101 98 · 82 91 67 50 32 c.c.

Experiment 6.—Well-nourished male, æt. 30. Knee very rigid. Pressure 52 centimètres.

Angle ´   .   0°   10° 20° 30°   40° 50° 60° 70° 80°

Volume   . 108½ 112½ 125 125½ 124½ 115 105 101 95 c.c.

The results which follow from these researches I may sum up in the following propositions :

1. That the knee-joint, in equal stages of flexion in different individuals, shows a very great difference in the capacity of its synovial membrane.

The difference of the pressure need not be taken into account, as, indeed, at the lowest pressure the volume of fluid in the joint was a maximum. It is the connection of the joint cavity with neighbouring synovial sacs which causes this phenomenon.

---

* I have left the figures referring to the volumes in cubic centimètres and the pressures in centimètres, since, for any practical purpose for which this table may be available, the following equations will facilitate their reduction to English measure:

1 centimètre = ·3937 inch = ·4 inch nearly.

1 cubic centimètre = ·061 cubic inch = ·0352 fl. oz. nearly—Tr.

2. That the capacity of the synovial cavity reaches its maximum in a definite degree of flexion, and that the angle at which this happens is 25°.

We learn from this that the statement of Bonnet, that the maximum capacity happens in the position of semi-flexion is incorrect, as we see that the position in which this condition exists is rather at the commencement of flexion.

But a second and not less interesting relation is evident from the preceding experiments. It is that the increase of capacity is the greatest from extreme extension to 10° of flexion, less from 10°—20°, and still less from 20°—30°. An important practical fact follows from this, that a slight degree of flexion, such as 10°, determines the relatively greatest increase of capacity of the capsule.

If the joint be in the position attained, when filled with fluid to its greatest extent, it may be forcibly extended without fear of rupture of the capsule; and here, again, my results differ from those of Bonnet.

3. The minimum of the capacity of the synovial cavity coincides with the maximum of flexion. Hence, the idea expressed by Bonnet on the method of treating penetrating wounds of joints is disproved—that the extension is the position in which the capacity of the capsule diminishes. Although in extension, as sections of frozen knee-joints show, the joint surfaces are closely approximated by means of the tensely stretched lateral ligaments, the spaciousness of the capsule in this position is, nevertheless, very considerable; and it is larger in semi-flexion than in complete. If the knee be forcibly flexed, and if the joint be now entirely filled with fluid, there ensues a degree of flexion by which the wall of the capsule is ruptured and the fluid escapes into the cellular tissue.

Moreover the clinical relations throw considerable doubt upon the correctness of Bonnet's theory of the mechanism of the knee-joint. In such cases as acute arthro-synovitis, the ligamentous structures specially suffer, and disease is distinguished by copious effusion into the articulation. We frequently find complete extension of the knee-joint throughout the course of the disease—an observation which I have repeatedly made, and

PLATE XIX   155

which is corroborated by Volkmann ('Krankheiten der Bewegungsorgane,' 1865, p. 195). Again, effusion of blood into the joint in an extended position of the extremity exhibits symptoms compatible with this.

Figure 2.—This section of a normal right foot is from the same body. The section runs near the inner border of the foot, and divides in succession the tibia, astragalus, scaphoid, internal cuneiform and first metatarsal bones, and the first phalanx of the great toe. The saw has missed the second phalanx, as the toe was somewhat bent outwards.

The section passes nearer the inner border of the foot than that represented by Weber ('Gehwerkzeuge,' tab. xi), Volz ('Beitrag zur Chirurg Anat.,' tab. x), Henle ('Gelenke,' figs. 136, 137). It was only just possible to avoid the cuboid and third cuneiform bones which project inwards so much that they would have been divided by any section passing further outwards, and made the relations of the plate more complicated. The bones of the foot are not placed so that they simply form an arch from before backwards, but there is also one in a transverse direction.

It can be easily proved by measurement, that from the pressure exerted by the weight of the body, in the upright position, the curves of the skeleton of the foot are flattened in both directions, and that the foot is not only lengthened but broadened.

It is clearly seen from the plate, that the astragalus which has been divided exactly at the attachment of the interosseous ligament, is set as the keystone of the arch. It is wedged in between the scaphoid and os calcis, is pressed against them both, and thus prevents their approach.

The ligaments correspond with the structure of the arch, which the several bones of the foot form. They are proportionally weaker on the convex dorsum, where they hold the separate bones in position during pressure on the arch; and extraordinarily strong on the plantar aspect, where their function is to act as a tie beam, and prevent separation of the bones: and it is not the form of the bones alone that renders the arch secure, since they would fall apart were it not for the immensely strong ligamentous arrangement of the sole of the foot, strengthened by the plantar fascia.

There is no necessity for mentioning the individual parts. The accurate drawing itself sufficiently explains the soft parts. Some notice must be taken of the pad of fat which is so largely developed at the point of greatest pressure on the sole, and which diminishes and distributes as much as possible this pressure over different points. Over the heel and in the region of the ball of the great toe it is half an inch thick; thus affording a soft support, which partially equalises the irregularities of the ground.

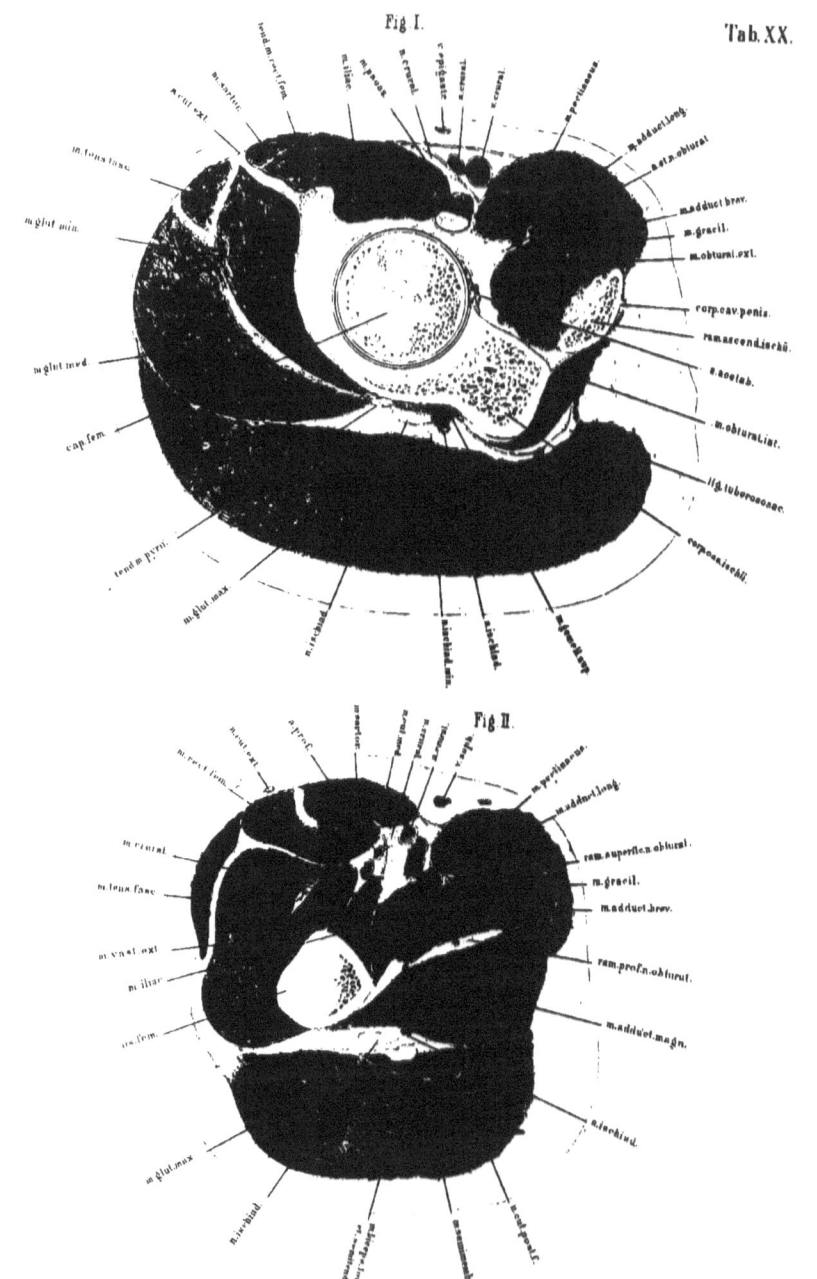

Fig I.

Tab. XX.

Fig II.

PLATE XX

THESE two sections of the thigh were taken from the same individual as Plates I A and I B. The sections were so directed that the first (tab. xx, fig. 1) passed immediately below Poupart's ligament and parallel with it, but obliquely with the direction of the thigh itself; it is consequently a section of Scarpa's triangle, and should be compared with that given by Legendre ('Anat. Homolograph,' Pl. XXIII), and by Voltz ('Chirurg. Anat. der Extrem.,' Tafl. vi, fig. 3). The second section (tab. xx, fig. 2) was not parallel with the first, but at right angles to the axis of the thigh near the perineum, so that the two sections would include a wedge taken out of the thigh, with the base external and the apex internal.

The following sections ran parallel to each other, and they form a segment of about 1·6 inches thick. They are from a very muscular thigh, and form a series. The other sections, from below the knee to the foot, are taken from another, though equally normal, male subject, and show the same relations.

The upper surfaces furnish the plates; and these, from the symmetrical structure of the extremities, will serve equally well for either limb, although they happen to be taken from the left; by being reversed they will correspond with the right, so that the under surface may be regarded as the stump of an amputation.

With regard to the bones, we first notice, in tab. xx, fig. 1, the absolutely circular section of the head of the femur completely surrounded by a thin layer of cartilage, behind which is seen the cavity of the joint as a dark circle. It is enclosed by a portion of the acetabulum, which is joined by the divided part of the ischium, or rather by its upper ramus. The section has then passed through the obturator foramen, obliquely outwards through the ascending ramus of the pubis and corpus cavernosum penis, the obturator membrane, and the sacro-sciatic ligament.

Above the capsule of the hip-joint through the divided synovial membrane is observed the psoas muscle and the portion of the iliacus associated with it. Below the outer extremity of this muscle is the section of the tendon of the rectus femoris. The second head of this muscle is incorporated with the ligamentous structures at the brim of the acetabulum, and could not be shown separately in the plate.

Above the ilio-psoas, is seen the fascia over the last dorsal nerve running down over the vessels to unite with the fascia of the pectineus, and attaching itself to the capsule of the hip-joint. We have here already the commencement of the sheath of the femoral vessels, and observe how it forms a prismatic space, the outer wall of which bears towards the sartorius. The superior boundary of this space would be indicated by a single lamina, as is shown in the preparation. External to the sartorius is the origin of the tensor vaginæ femoris attached to its tendinous sheath, and between them the external cutaneous nerve. Next, we observe the gluteus medius muscle with its strong tendinous fascia from which a portion of its fibres arise. The oblique section of its muscular bundles is not quite clearly rendered in the plate—a remark which also applies to the gluteus minimus, which is more internal. To the latter is attached the tendon of the pyriformis, and of the gemellus superior and obturator internus, which is seen in its angular course with its large subjacent bursa.

The above-mentioned series of muscles forms the superior limit of the space occupied by the vessels and nerves, as the gluteus maximus does the inferior. The great sciatic nerve is here seen. The fascia which comes from the gluteus medius, to cover the gluteus maximus, is considerably thinner on the latter muscle, passes over this median ridge to be partly inserted into the great sacro-sciatic ligament, and partly into the fascia of the obturator internus.

Of the adductor group are seen the sections of the pectineus, the adductor longus, and the adductor brevis. The adductor magnus is not seen; and the

PLATE XX 159

gracilis is divided in its tendinous origin. The acetabular artery, which in this case comes from the internal circumflex, lies close on the hip-joint. Care has been taken to represent the direction of the fibres of the muscles, and also the masses of the several bundles of fibres as accurately as possible, the coarse fibres of the gluteus maximus being particularly noticeable. It is true that from this plate hardly sufficient can be gathered to form a correct idea of the formation of the crural ring, and the anatomical relations of crural hernia; but we shall have to rest contented with having obtained the idea of the size of the individual portions and the position of their layers with regard to each other in their natural relations, and I do not think that we should have gained more if the section had been taken farther up. Linhart has already correctly remarked, that for the representation of the relations of crural hernia single sections are not sufficient.

Plate XX, fig. 2, is a section of the thigh at right angles to its axis immediately below the trochanter minor. The lower portion of the iliacus muscle is still seen on the inner surface of the thigh; close to it and internally the pectineus; and externally the crureus. The femoral artery has already given off the profunda, which is separated from the main trunk by a lamina of fascia.

The three adductors lie over one another on the inner side; and above and beneath the adductor brevis are the two branches of the obturator nerve, with it the branches of the internal circumflex artery. More internally is the gracilis, which is now fleshy.

The sartorius is drawn more over to the middle, and is on the point of overlapping, like a muscular roof, the femoral artery, which vessel has acquired a more superficial position with respect to its accompanying vein.

The rectus femoris with its internal tendinous raphé, lies on the crureus and vastus externus, and near it the tensor vaginæ femoris, which is enclosed by the fascia common to it and to the tendon of the gluteus maximus.

The strong lamina of fascia which passes beneath from the gluteus maximus, and turns inwards between the vastus externus and rectus femoris, is worthy of notice. The tendon of attachment of the gluteus maximus to the bone is not yet seen, but its insertion into the fascia lata only, which is especially developed at the external surface of the thigh. Covered over, but separated from it by a thin lamina of fascia, is the common head of the biceps and semi-tendinosus, and above that the strong tendon of the semi-membranosus. Between it and the adductor magnus is the great sciatic nerve, and a large inosculating branch of the ischiatic artery, with the first perforating and the profunda.

The segment, the upper surface of which is here represented, was about two inches thick. If the sections of the arteries in both plates be compared, it will be seen that the femoral artery changes its position with regard to the bone, and to its accompanying vein, in its course downwards. At the level of the horizontal ramus of the pubes it lies so near the bone, that the possibility of its compression against it was obvious; in fig. 1 the distance of the artery from the head of the bone is so inconsiderable that pressure could be readily exerted on the vessel; whilst in fig. 2 greater pressure would appear to be necessary.

Besides this distance of the vessel from the bone there is also an alteration in its direction. In figure 1 the artery lies above the bone, so that a force acting vertically from the front might smash both bone and vessel; in fig. 2 it lies farther down, already so far to the side of the femur, taking its course outwards, that a force acting in the same direction might wound the artery without injuring the bone, or the contrary.

Moreover the position of the artery to the vein changes during its course. Commencing at the abdominal cavity, the main trunks lie alternately in the sagittal and frontal planes. The abdominal aorta lies on the lumbar vertebræ close to the vena cava. In the abdominal cavity the iliac artery lies in front of its vein, at the inner border of the psoas; and then, after passing below the crural arch, lies to the side of the vein. The vessels, however, soon again change their relation, for below the fossa ovalis, as is seen in fig. 2, the vein lies below the artery and accompanies it to the knee; so that, in attempting to reach the popliteal artery from behind, the vein would be in danger of being wounded, and must be pushed aside in order to render the artery accessible.

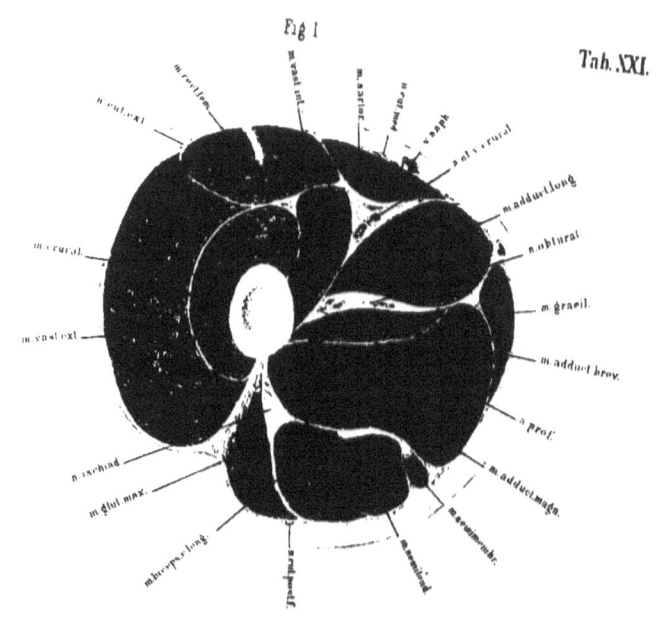

Fig. I

Tab. XXI.

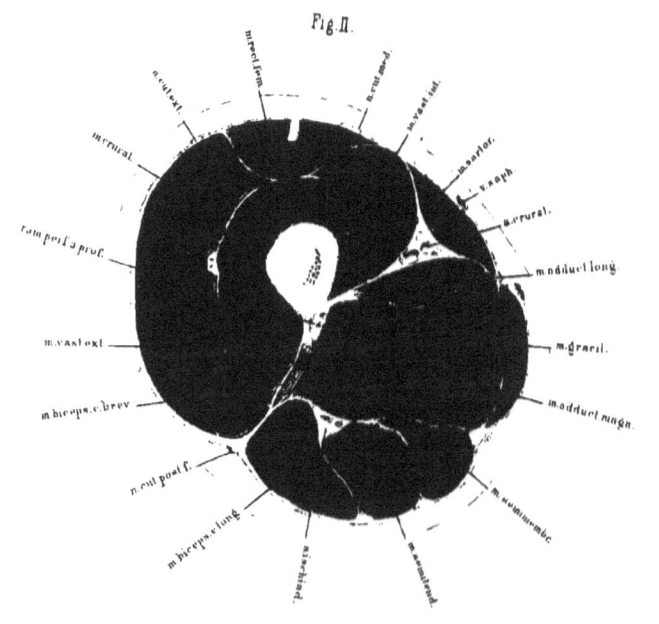

Fig. II.

PLATE XXI

Fig. 1 of this plate is a section of the thigh taken somewhat below the upper third, about 2·5 inches below the section shown in the preceding plate, and three inches below the trochanter minor.

The individual portions of the quadriceps extensor are clearly seen separated from each other by fascia. At the posterior border of the vastus externus, which is covered by the strong dense fascial tendon of the tensor vaginæ femoris, is the termination of the gluteus maximus. This muscle is attached by means of a strong tendinous mass to the thigh bone, and here separates the flexor muscles from the extensors.

Of the flexors which accompany the ischiatic nerve the biceps and semitendinosus are now completely separate.

The semimembranosus has already become muscular. Over it lie the three adductors—first, the adductor magnus; upon it the adductor brevis; and between this and the adductor longus the profunda artery and obturator nerve.

On the other side of the adductor longus, between it and the vastus internus, is the space for the femoral artery and vein. The sheath of the vessel is clearly seen ; its formation by fascial laminæ; and its closing-in by the sartorius, which continually approaches the inner side of the thigh. This muscle reaches the gracilis, to which it is very similar in form, getting closer and closer to it until at last the two muscles accompany each other.

Fig. 2 represents a section through the middle of the thigh, where the sartorius and gracilis meet, and the short head of the biceps begins to take the place of the gluteus in the external intermuscular ligament, between the vastus externus and the flexors. External to the rectus femoris the individual portions of the quadriceps are not seen any more, the rectus with its central tendon being completely isolated by fascia.

21

The femoral artery, which begins to lie considerably more laterally with regard to the bone, is still in the same fascial sheath, between the adductor longus and vastus internus, and covered by the sartorius. The adductor longus has already lost its bulk; and the adductor brevis has disappeared at this level entirely.

The profunda artery is divided at the point where it perforates the adductor magnus close to the bone.

The three flexor muscles are completely isolated from each other, and lie so close together posteriorly that the great sciatic nerve takes a position in a furrow between the long head of the biceps and the semi-tendinosus.

With reference to this plate, it may be added that the thigh was rotated somewhat outwards before the section was made.

Tab.XXII.

Fig I

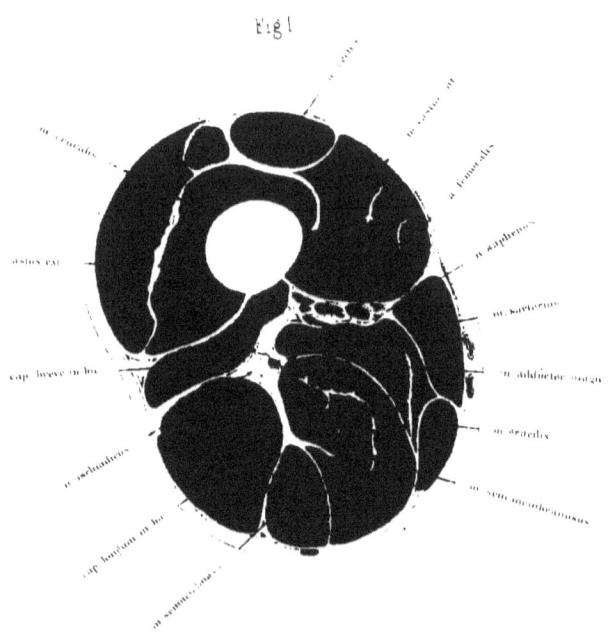

Fig II

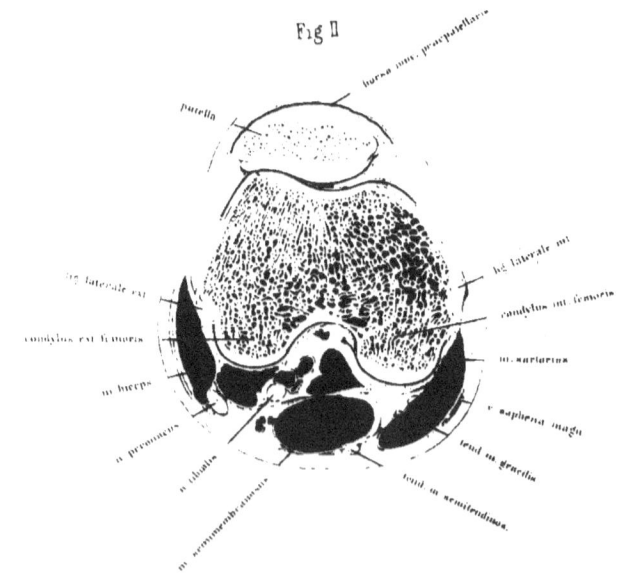

·PLATE XXII

THIS and the following plates are taken from sections from another body, but can be used equally well in the series. The arteries were injected, the body frozen lying on the back, and the lower extremities left in their normal position, *i. e.* somewhat rotated outwards.

· Fig. 1. The section here passed through the lower third of the thigh nearly a hand's breadth above the upper border of the knee, at the position of the passage of the artery through the adductor opening. The plate represents the left thigh, and the upper surface of the lower portion; the external aspect of the extremity being to the left, and the internal to the right.

The adductor longus is not seen, as it terminated just above the line of section. Of the adductors the magnus only is present; its section is associated with the great vessels. It is no longer attached to the linea aspera, but all the muscular tissue to be seen here passes directly into its tendon, which terminates at the internal condyle of the femur.

This is exactly the spot where the artery passes through the adductor opening, in order to reach the back of the bone. The artery itself lies surrounded by a system of veins, which render ligature a matter of difficulty, on account of their free anastomosis. Between the artery and the bone lies the vein, with two small ones opening into it; on the opposite side are two venæ comites, which are lodged between the artery and the long saphena nerve. If the artery be tied at this level, the incision must be made between the sartorius and the internal vastus, but upon the outer side of the former; the strong dense fascia under the sartorius must be divided; and the saphena nerve and venæ comites pulled on one side; there is thus more difficulty in reaching and isolating the artery in this place than higher up (*vide* Plates XX and XXI).

It is not correct to describe the course of the artery as spiral with regard to the bone. It lies certainly in front of the bone above, in the middle of it farther down, and at the knee-joint completely behind it. One can convince oneself on any preparation, whether the artery be injected or not, that the artery passes downwards in a tolerably straight direction; it is the bone on the contrary that describes a twist round the artery. The relation of the artery to the sartorius is constant throughout the entire length of the thigh.

·The great sciatic nerve, like the artery, has changed its position from the upper section. As higher up it lay behind the adductor magnus, so here it will be seen behind the short head of the biceps. There is nothing further to say about the muscles. The prominence of the central tendinous intersections indicates the termination of the muscles, as well as forming the separation between the individual portions of the quadriceps, which higher up were separated by fascia.

Fig. 2 is a section of the left knee-joint through the centre of the patella. The man, whose lower extremity afforded the preparation, had been a mason, and had probably knelt a great deal. At all events the large development and width of the præpatellar bursa would suggest it. The patella lies with the external portion of its posterior articular surface so close to the external condyle of the femur, that only a narrow chink separates them; while on the other hand it is raised from off the external condyle. The synovial cavity is divided by means of the ligamentum mucosum into two portions; of these, one follows the patellar surface and passes upwards and inwards, whilst the other is applied over the inner condyle. This position of the patella upon the condyles renders it clear why in dislocation it glides by preference over the external condyle. The position itself is conditional on the curving inwards of the femur, so that the action of the powerful extensor muscles alone would cause the patella to glide outwards from off the flat hollow between the condyles, if these lateral tendinous masses did not securely hold it in position. These structures are interwoven as fibrous bundles with the lateral flat tendinous expansions which pass from the great extensor downwards to the leg, and assist in transmitting the power of extension beyond the patella and ligamentum patellæ. In fracture

PLATE XXII   165

of the patella they keep the fragments in position; and, if the fracture of the bone be transverse, they are torn simultaneously with it. If the patella be sawn through, on the body, maintaining, however, these lateral ligamentous structures, and the leg be flexed, the halves of the patella separate slightly from each other; if, however, they be divided in addition, there immediately ensues a very wide separation of the fragments. We can thus understand why stellate fractures of the patella unite by bone, as in this instance the patella alone is involved; whereas in transverse fracture the ligaments are also torn, the extensor muscles dislocating the upper fragment.

The plate shows broad ligamentous bands passing from the patella to both sides of the femur, and surrounding the entire knee-joint anteriorly and laterally.

Behind the articular surfaces of the condyles is the expansion of the synovial cavity between the crucial ligaments. The nerve, artery, and vein, lie close behind each other, the former being more external; the sciatic nerve dividing into the external popliteal inside and below the biceps, and the internal popliteal more towards the middle.

The muscles, which in fig. 1 showed such fleshy masses, are here confined and diminished in bulk. They are for the most part completely reduced to tendon; and the defined form of contour, which is characteristic of the region of the knee-joint, is dependent on that of the bones.

In synovitis, the patella would be lifted off the articular surface of the thigh bone, the distension of the capsule being especially evident in front. The posterior parts are but slightly yielding, and are consequently only slightly separated from the posterior surfaces of the condyles.

PLATE XXIII

Fɪɢ. 1 is a section through the upper third of the leg taken from the same subject as the last.

Similar plates will be found in Volz (*a a O*, taf. ix, fig. 1) and in Pirógoff (fasc. 4, tab. viii, fig. 8).

The strong framework from which the muscles spring is formed by the tibia and fibula, the inter-osseous membrane, the strong fibular intermuscular aponeurosis, which passes obliquely outwards and forwards from the fibula between the peroneal and extensor muscles, and the dense fascia, from which the fibres of the tibialis anticus in particular arise.

The strongly developed muscles divide themselves into three groups. Anteriorly are the extensors, tibialis anticus, and extensor communis digitorum bounded behind by the interosseous membrane, the extensor longus pollicis is not yet seen, as it arises lower down. Externally and lying on the fibula is the peroneus longus, which belongs to the second group separated from the extensor communis by the intermuscular ligament. Posteriorly to both bones is the third group, in which the flexors preponderate, and their deep layer is analogous to that of the extensor side in having only two muscles.

The tibialis posticus lies on the interosseous ligament between the tibia and fibula, and the flexor longus digitorum, of which only a small portion is seen, on the tibia. Behind them are the large expanded surfaces of the soleus and gastrocnemius, and on the posterior aspect of the tibia is a strip of the popliteus. This muscle lies almost entirely between this and the last section (Plate XXII).

The nerves belonging to the three groups are marked white. The superficial peroneal nerve lies between the peroneus longus and the fibula; and the deep peroneal nerve, which is separated from it by the fibular inter-

Tab. XXIII.

Fig.I

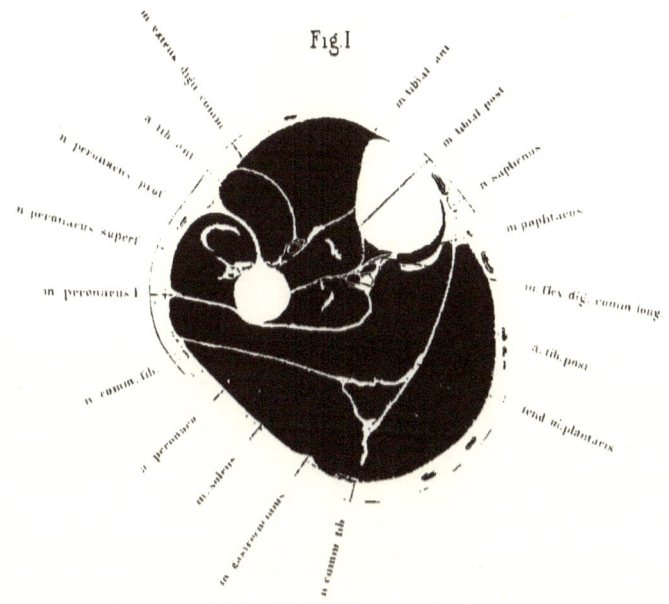

Fig.II.

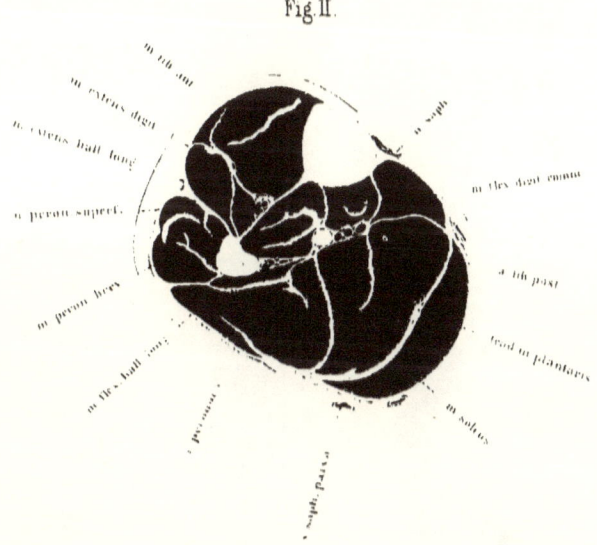

PLATE XXIII 167

muscular septum lies on the interosseous ligament and fibula. The posterior tibial nerve is seen between the flexor longus pollicis and the soleus.

The three arteries, the anterior tibial, posterior tibial and peroneal, are seen together with their veins. The two latter arteries lie close to each other, as the section passed immediately below their origins, separated from the interosseous ligament by the tibialis posticus, and they divide the deep layer of the flexor group from the muscles of the calf which form the superior layer. The anterior tibial artery lies on the interosseous membrane. The furrow running between the tibialis anticus and extensor digitorum longus indicates the position of this vessel, hence it may be readily found, its depth being the only difficulty.

Fig. 2. This section through the middle of the left leg may be compared with the plates of Henle ('Muskellehre,' fig. 142), and Voltz (a a O, tab. ix, fig. 2).

The relations of the muscles, vessels and nerves can be so readily made out that it does not seem worth while explaining the plate.

Beneath the muscles of the calf, in this section, all the flexors are seen together. The flexor longus digitorum has now considerable bulk, and so also has the flexor longus pollicis, which has already the peroneal artery between it and the fibula; and the anterior tibial artery lies between the extensor communis and the tibialis anticus. The artery is still so deep that its ligature at this place, though practicable, is not to be recommended. Farther down, and nearer the ankle, the muscular tissue ceases somewhat, and the vessel is more easily reached.

The peronei muscles are completely developed, and the superficial peroneal nerve is already approaching so near the surface that it seems about to perforate the fascia.

If the two sections be compared which represent the position of the individual structures in the upper half of the leg, the superficial position of the tibia is evident and can be readily felt, hence affections of this bone from disease and accident are easy of diagnosis. The fibula, on the other hand, is unfortunately situated in this respect. The thick masses of the surrounding muscles do not favour its examination, and we must in its instance use some other diagnostic means, such as fixed, deep-seated pain.

The course the knife must take in order to expose the fibula is · indicated by the fibular intermuscular septum. The muscles which bound this septum, the peroneus brevis and extensor proprius pollicis, are easily made out at the outer surface of the bone, and the plate assists the surgeon in judging of the depth the wound should be in muscular individuals. In this proceeding no vessels of large size will be met with, but the peroneal nerve must be carefully avoided, as it would fall in the line of incision.

The superficial position of the tibia also demands attention with regard to the treatment of ulcers, as the periosteum is all the more likely to be involved the fat being so sparingly developed, whilst in almost all other points of the section it is more abundant and consequently the skin is further from the subjacent fasciæ.

The main arteries, with their accompanying veins, at this level are still tolerably near their origins, and not very far separated from each other. In the inner portion of the section they lie so near the interosseous ligament, and are so protected from external pressure by the bones of the leg, that they are not so liable to be wounded as elsewhere. At the same time from their position they are not readily compressible against the skeleton, so that in amputation or any operation where much bleeding is expected a tourniquet must be applied above the knee.

Fig I.

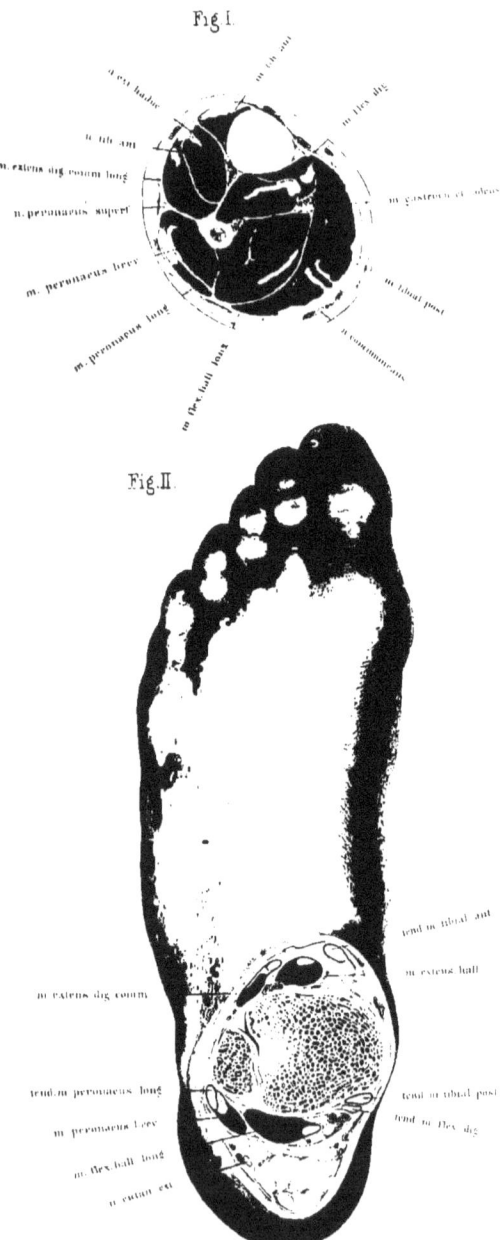

m ext hallue   m ext ant   m flex dig

n tib ant

m extens dig comm long

n peronaeus super   m gastrocn et soleus

m peronaeus brev

m peronaeus long   m tibial post

m flex hall long   n communicans

Fig.II.

tend m tibial ant

m extens hall

m extens dig comm

tend m peronaeus long   tend m tibial post

m peronaeus brev   tend m flex dig

m flex hall long

n cutan ext

PLATE XXIV

Fig. 1 represents a section through the lower third of the left leg near the joint. From the decrease in the masses of the muscles and the increase of the tendinous structures the section of the limb has become considerably smaller. Although individual muscles, such as the extensor and flexor longus pollicis with the peroneus brevis, have become stronger than in the preceding plate, they do not make up for the want of those of the calf which determine the size and shape of the leg. The soleus and gastrocnemius are no longer separate, a longitudinally directed tendinous mass spreads over the posterior surface of the soleus; this is the termination of the gastrocnemius, which becoming blended with the fibres of the soleus, forms the tendo Achillis.

The largest surface shown is that of the flexor longus pollicis, which is here divided at its greatest bulk. In flexing the great toe in walking this muscle contracts so forcibly that its power exceeds that of the other flexors of the toes. Its position has altered from the last plate, being further back and more beneath the tibialis posticus, so that after completely crossing it in the malleolar region it lies at last most internally.

The position of the deep flexors is essentially distinct from that of the extensors. The tibialis anticus lies close on the tibia, and gains the inner border of the foot without crossing its neighbours, the extensor longus pollicis and extensor communis digitorum; whilst the tibialis posticus lies in the middle on the interosseous ligament, the flexor longus pollicis on the fibula, and the flexor longus digitorum on the tibia, and these muscles cross each other before their ultimate insertion. This position is connected with their passage at the inner malleolus. As they are

22

pushed aside by the sustentaculum ·tali, they would obtain a very insufficient hold beneath the short internal malleolus if the flexor longus
pollicis and tibialis posticus lay on the inner border of the leg, and if
the flexor longus digitorum arose from the fibula it would act at a great
disadvantage. This defect is remedied in a simple manner by the crossing
of the tendons.

The arteries have the same muscular separations as before, notwithstanding that they have materially altered their position with regard to the
tibia; and, in consequence of the diminution of the bulk of the overlying
muscles they are considerably nearer the surface, so that their ligature is
easier than above. The anterior tibial artery can be reached between the
tibialis anticus and extensor longus pollicis, and the posterior tibial can be
readily found if the border of the soleus be detached and pulled back
from the flexor longus digitorum. The position of the peroneal artery is
the most unfavorable for ligature, as it must be searched for behind the
peronei, after separating the flexor longus pollicis from the fibula, when it
can be drawn out from behind the bone.

Fig. 2. This section of the leg in the region of the malleolus terminates
this series. It divides the tibia and fibula immediately above the astragalus,
hence the comparatively large size of the tibia. Both are strongly bound
together by ligaments, and in front is an opening into the cavity of the
ankle-joint.

The muscles now almost entirely present their tendons, only the outer
portion of the extensor of the toes, the peroneus tertius, and the extensor
flexor longus pollicis, still show muscular tissue. With the tendons are
associated their bursæ which are shown as dark chinks, and the ligamentous
apparatus which renders secure the position of these tendons at the ankle.
The upper portion of the annular ligament is met with, the point of origin
of which from the os calcis lies deeper and is consequently not seen,
and under the middle fasciculus which encloses the extensor longus pollicis,
lies the anterior tibial artery which may be here readily reached from the
surface. To expose the posterior tibial artery for ligature, the division
of one fasciculus only of the internal annular ligament is necessary. It
lies between the flexor longus digitorum and flexor longus pollicis, and the

PLATE XXIV 171

bursal sheaths of both muscles can be completely avoided in looking for the artery. The tendo Achillis lies some way further back, so that its division is easily accomplished without wounding the vessel.

The two plates here given are sufficient to show the most important points in the lower half of the leg. On the other hand, the relations given of the foot are insufficient, and perhaps a further series of sections might have been shown. From numbers of sections which I have made and had drawn, and have before me, as well also from the examination of Pirogoff's plates, I have come to the conclusion, that sections of the foot are not of very much use for the comprehension of its structure, although a clear idea of the arrangement and form of its bony arches may be obtained; but for the relations of the soft parts they are only of subordinate importance. Flat preparations are in this respect of more value and are indispensable. The numerous small muscular masses of the sole are divided from each other merely by fasciæ and cellular tissue, and the number of tendons on the dorsum which can be but inadequately separated from the ligaments by transverse section, would give unreliable plates. Again, the arrangement of the annular ligament would be absolutely unintelligible if studied on sections only. The arteries, as has already been mentioned in fig. 1, lie much nearer the surface than in the preceding plate, and therefore have far simpler landmarks for their ligature than in the upper half of the leg. They form a triangle with two nearly equal sides. The base of this triangle is formed by a line passing from the anterior tibial artery to the peroneal, directed outwards, as seen in fig. 1. This arterial triangle, in consequence of the termination of the peroneal artery, ceases in fig. 2, and is not seen in Plate XXIII, fig. 1. On the other hand, it is very clear from Plate XXIII, fig. 2, that if this triangle be compared in this and the preceding plate, the direction of its base and the length of its sides remain exactly the same. It so happens that these arteries in their course in the lower half of the leg remain in the same position with regard to each other; and that they run as parallel vascular tubes, and do not from their own change of position get nearer the surface, but from the continually decreasing masses of the muscles as they proceed downwards.

PLATE XXV

THE accompanying frontal section of the thorax and shoulder-joints was made from the body of a very powerful man. Beyond the enlarged thyroid body there was nothing abnormal. From the recumbent position of the body, particular regard was taken of the upper extremity, and it appeared desirable to divide the humeri in their long axes, and the arms being placed in the position they would have held in the upright position were rolled outwards so that the bicipital groove was directed forwards. After being frozen in this position, the head was removed from the neck just below the larynx, and the rest of the body separated by a section through the nipples. The frontal section was so directed that it passed through the middle of the heads of the humeri and their shafts.

Before freezing, the arteries were injected from the femoral.

The cupolæ of the lungs are divided through their highest points. Both subclavian arteries pass over the cupolæ of the lungs, and consequently cause an impression on the pleura, which on examining the cavity of the chest can be readily recognised.

The arteries, however, do not cross the cupolæ of the lungs at their highest points. They lie considerably behind them and below the brachial plexus in the neighbourhood of the head of the first rib. The section has passed through the arch of the right subclavian artery, but not disturbed the left, running in front of it as is clearly seen in the plate. The preparation showed on further examination that the lungs and pleural cavities extended considerably further up. The first ribs were divided at their anterior extremities, the right behind the origin of the scalenus anticus, the left immediately through its origin.

The roots of the lungs lie behind the section, the left further from its plane than the right.

Tab. XXV.

PLATE XXV   173

Corresponding with this, on the left side of the plate, there is no inter-
ruption of the pleura, whilst on the right side (to the left of the spectator),
the points of reflexion of this membrane have fallen in the section. The
relations are complicated by the pericardium. Between the lungs and
heart there are seen two spaces, which are the cavities of the pericardium
and pleuræ.

The left ventricle is opened, and a portion of the right auricle is shown.
In connection with them are seen the aorta and superior vena cava in
section. The former is exposed for its whole extent, so that the entrance
from behind of the azygos major vein appears. In continuation of the
superior cava is the right innominate vein, which as it passes more vertically, is divided throughout, and the two delicate valves are seen. The left
innominate vein, which passes more obliquely, was removed with the other
half of the body. Its end only is shown, at the point of entrance of the left
subclavian vein as a large venous lumen immediately above the first rib.

The aorta is exposed in the horizontal portion of its arch. At its
origin it shows a considerable swelling of the bulbus aortæ, produced by
the pressure of the injection on the semilunar valves, of which two, one
nearly bisected, are seen. Below them, in the left ventricle, is the aortic
segment of the mitral valve. The liquor pericardii had collected in the
upper portion of the pericardium.

It will be observed from the free surface afforded by the divided
left auricular appendix above the left ventricle, that the two laminæ
of the pericardium are considerably separated from each other in this
situation, whilst in all other places they are directly in apposition, so that
its cavity is shown only as a crevice. Between the left ventricle and
the ascending aorta is the section of the pulmonary artery, which being
nearly horizontal, is divided transversely. The vessel is seen from before
backwards, and the lumen of the right branch is exposed, curving sharply
behind the aorta, to reach the root of the right lung; whilst the left branch
passes obliquely upwards and outwards, to course over the left bronchus
and root of the left lung.

The position of the great vessels given off from the aorta is considerably
altered by the hypertrophied thyroid gland. This, as the plate shows, has

compressed the trachea on both sides; and very probably interfered with deglutition from pressure on the œsophagus. It involved the interspace that the two carotids form with the aorta, and pushed them asunder. In the left carotid, which is freely divided, this is clearly seen; whilst in the right a small portion only of its origin from the innominate is involved, as it lay almost entirely in the anterior half of the preparation.

The subclavian artery of the left side is not seen, as it takes its origin from the arch of the aorta behind the carotid; it lay in this preparation behind the section, covered by the pectoralis minor. Its continuation, the brachial artery, came into the line of section, and is to be seen between its accompanying nerves.

On the right side is seen, on the other hand, the continuation of the innominate artery into subclavian and axillary. The arch of the right sub-clavian passes under the right innominate vein, over the cupola of the right lung; and gives off anteriorly the internal mammary artery, which is here transversely divided, and the inferior thyroid which is slit up and covered at its extremity by the thyroid body; passes over the first rib from within outwards; and finally disappears behind the cut surface of the coraco-brachialis.

The subclavian veins correspond on both sides. The right subclavian vein is cut short off above the second rib, and the left is widely opened between the scalenus anticus and pectoralis minor. The latter, which receives many small veins, is of large calibre, and passes with its inner wall rather more upwards, towards the internal jugular vein which lies on the outer side of the carotid artery. Of the internal jugular vein of the right side nothing is to be seen, the parts being entirely removed with the anterior half of the body. The left subclavian vein consequently lies farther forward than the right.

The right brachial plexus is exposed throughout its length, whilst the left is covered and only its commencement is seen under the anterior scalene muscle.

The several structures of the neck group themselves about the fifth, sixth, and seventh cervical vertebræ. At the lower border of the seventh cervical are the cut surfaces of the longi colli muscles, which lie between

PLATE XXV   175

the spine and the thyroid gland.   Above both muscles, on either side of the
bodies of the vertebræ, are the vertebral arteries slit open ; of these the left
shows a far larger calibre than the right.   From behind these vessels
proceed the roots of the brachial plexus, which is entirely covered on the
left side, and partly on the right, by the cut surfaces of the scaleni.   Still
more externally and upwards are the sections of the sterno-cleido-mastoids,
with a strip of the platysma, immediately beneath which on both sides is
the external jugular vein.

   The right phrenic nerve is completely removed ; the left is seen between
the carotid artery and the lung.   The artery accompanying it is the internal
mammary.

   The vagus is only partially cut on the left side, where it lies in front
of the arch of the aorta, and from whence its recurrent branch passes
upwards behind that vessel.   On the right side, on the contrary, it is
divided transversely at the point where it is applied to the root of the
lung.

   The shoulder-joints have so fallen into the section that the saw has passed
on both sides in front of the glenoid cavities ; and nothing is seen of the
scapular element of these articulations.   The bony elements of this portion of
the joint lie behind the plane which passes through the middle point of the
head of the humerus.   On the left side the glenoid cavity was only a quarter
of an inch behind the plane of section ; on the right it was so much closer that
the limbus cartilagineus fell into it.   As the head of the humerus is directed
inwards and backwards towards the glenoid cavity and as the section passes
deeper on the right than on the left, the greater tuberosity of the right
side has been entirely removed.   The round section of the head is all that
is seen, whereas on the left the greater tuberosity projects in a triangular
form.

   On the right side a portion of the acromion appears ; and on the left the
section has passed more anteriorly, and has nearly divided the coraco-
acromial ligament.   Normally the acromion rises but very little above the
head of the humerus, so that anteriorly a tolerably large portion of the latter
remains unprotected by bony covering.   The coracoid process is divided
transversely on either side, and is readily seen between the head of the

humerus and the clavicle. It is cut through behind the attachments of the muscles.

The pectoralis minor on both sides of the chest is divided, and shows a large surface of section, on the left side particularly. This is explained by the forward position of the shoulder, and by the muscle becoming relaxed and folded so that its posterior border was bent backwards.

The strongly curved clavicle has a different appearance on the two sides. The right, which projects further forwards, shows beyond the section its entire acromial end, whereas on the left side this is not seen. The section of the clavicular portion of the deltoid of this side is shown. On the right side the anterior attachment of this muscle is completely removed. Its attachment to the humerus is equally divided on both sides; and the bursa between it and the capsular ligament appears as a black line.

With regard to the relations of this capsule, the following points are to be noticed. Since the shoulder-joint is under the influence of atmospheric pressure, the bone is pressed against the glenoid cavity; and therefore the cavity of the joint notwithstanding its extent and the laxity of its capsule can be shown merely as a crevice in the representation of its section. The ligamentous tissue which terminates at the neck of the humerus is the capsule: this, on the left side, encircles the bone like a ring from the greater tuberosity, and encloses the obliquely divided tendon of the biceps superiorly; these relations on the right side are shown rather differently. In the first place, a portion of the limbus carti-lagineus is seen, terminating above in a sharp angle, and externally the supra-spinatus presents itself in section strengthening the capsule by its tendon, and which more externally is so closely united with the tendons of the infra-spinatus and the teres minor that no line of separation can be represented.

On the inner side of the neck the capsule is more loosely attached, so that by raising the humerus its folds are obliterated.

The limit of the capsule towards the middle line is formed by the sub-scapularis, which is seen divided on both sides. Beneath it lies its bursa, which must be looked for between it and the capsule. It normally forms a communication with the cavity of the joint, but which was not seen in this

PLATE XXV   177

section.   Nevertheless the outer side of the subscapularis is to be seen on
the left shoulder-joint limited by a dark line, indicating the synovial membrane in section.   This line runs in a curved direction with its concavity
outwards, corresponding with the head of the humerus.

In order to demonstrate the extent of the cavity of the capsular ligament and to show the amount of separation of the humerus from the
scapula when the joint is distended by effusion, I injected some fresh
joints with tallow, froze them, and then made sections.   One of these preparations is shown in the following woodcut.

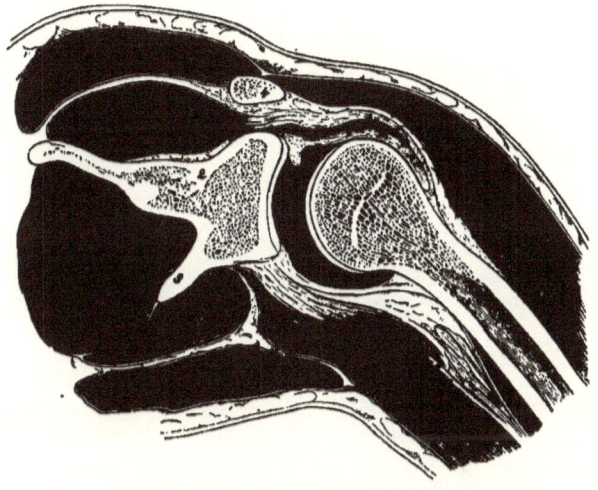

Frontal section of the right shoulder-joint, injected with tallow.   Anterior half.   ½.
1. Head of humerus.   2. Neck of scapula.   3. Anterior margin of scapula.   4. Clavicle.
5. Deltoid.   6. Triceps.   7. Teres major.   8. Teres minor.   9. Infra-spinatus.
10. Supra-spinatus.   11. Trapezius.

The humerus is seen from behind half extended and somewhat rolled
inwards, a position it acquired from the great pressure of the injection, and
corresponding with the greatest amount of distension of the capsule.   This
injection was made from the supra-spinous fossa through the glenoid cavity,
and the upper arm amputated at its lower end, so as not to hamper the movements of the joint by its weight.   It appeared that the greatest distance of the

23

head of the humerus from the glenoid cavity was somewhat over half an inch; hence it would appear likely that in inflammation with effusion into the cavity of the joint, there would be some considerable lengthening of the limb.

In order to bring the relations of the heart more completely into notice, it became necessary to extend the section farther downwards than was possible in this preparation. Consequently I made a series of sections to supply this deficiency, but unfortunately none of these specimens could be used to supplement this plate.

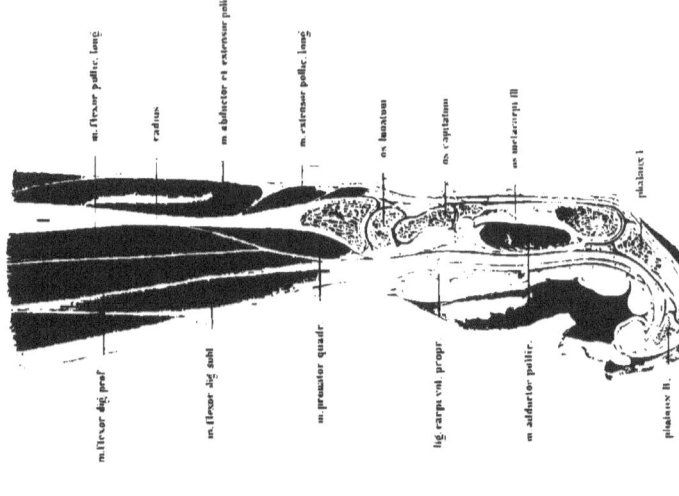

Fig II

m. flexor pollic. long.

radius

m abductor et extensor pollic brev

m extensor pollic. long

os lunatum

m. capitalum

os metacarpi III

phalanx I

m. flexor dig. prof.

m flexor dig subl

m pronator quadr

lig. carpi vol. propr

m adductor pollic.

phalanx II

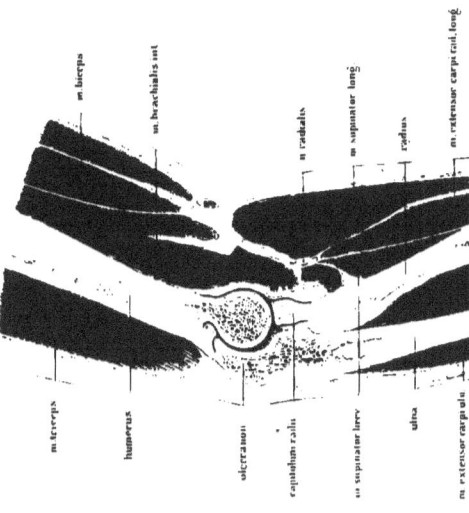

Fig. I

m. biceps

m. brachialis int

m radialis

m supinator long

radius

m extensor carpi rad. long

m triceps

humerus

olecranon

capitulum radii

m supinator brev

ulna

m extensor carpi uln.

PLATE XXVI

THE longitudinal section shown in this plate is taken through the elbow-joint and hand of a young normal female subject, with no previous injection of the vessels.

There was neither artificial injection of the articulation nor any pre-determined position thereof. It was frozen and sawn through in its normal condition.

Figure 1. In this plate is shown the sagittal section of the right elbow-joint, taken somewhat obliquely, and seen from the radial aspect. The saw has passed nearly through its centre, and removed a small portion of the radial surface of the ulna. As the forearm is slightly bent, and in semipronation, the radius is met with in its long axis, a small portion of the ulnar aspect of its head only remaining. Farther down its shaft is divided obliquely, and the medullary cavity partially opened. In consequence of pronation the radius does not lie parallel with the ulna but crosses it, and is directed with its inferior extremity forwards.

The expansion of the cavity of the elbow-joint is worthy of notice in flexion and extension of the humerus. The folding-in of the capsular ligament in the posterior supra-trochlear fossa corresponds with the slight degree of flexion, and if this flexion be further increased, this folding-in would become eradicated and take place on the anterior aspect. The cavities above the trochlea are alternately filled at the end of flexion and extension, the capsule, however, being drawn away beforehand by certain muscles, viz. the brachialis anticus and biceps, so that it may not be included between the bones.

It can be seen from the plate that the bones do not lie completely in apposition. Injections of the elbow-joint with strong pressure show that it acquires the position of semiflexion, and that the fluid injected partially separates the joint-surfaces.

The terminations of the flexor muscles of the arm are not seen. The brachialis anticus, which lies close upon the capsule, is divided longitudinally, as is seen from the direction of its fibres; and the same remark applies to the biceps, its tendon is deeper down behind the radius, and can be exposed only by dissection. The tendinous mass shown in the plate, between the upper end of the radius and the ulna, is a portion of the tendon of the biceps; another portion of it belongs to the circular ligament of the ulna, which forms the means of checking excessive separation, and becomes broader in pronation. The triceps at the back of the humerus shows its complete connection with the olecranon. On the anterior surface of the biceps is the supinator brevis, and farther in front are portions of the supinator longus and the extensor carpi radialis longior the heads of which are removed with the external condyle. No vessels or nerves are seen in the plate, excepting an obliquely divided vein, a portion of the median-cephalic, and the radial nerve beneath the supinator longus. The main artery, with its accompanying veins and the median nerve, which pass down on the inner side of the arm and afterwards turn forwards on the bend of the elbow, lie concealed in the soft parts below the surface of the section.

Sections made as shown in this plate are rarely successful, and not easy to understand at first sight, since in complete supination and parallelism of the bones of the forearm—the usual position from which descriptions are made—the radius and ulna lie in a frontal plane.

I have, however, specially chosen the present position of the arm for the section as being the more normal one in which the radius lies in front of the ulna for almost its entire length. A very similar representation is to be found in Pirogoff's Atlas (Fasc. iv B, Taf. iv, fig. 7).

Frontal sections of the elbow-joint agree with the preceding if the forearm is completely extended and supinated, and if, moreover, it be forcibly retained in this position before freezing.

The radius and ulna are divided in their longitudinal axes and in con-

PLATE XXVI   181

tinuation with the humerus. As Pirogoff's and Voltz's Atlases contain an excellent and complete series of such sections, it seems hardly worth while to multiply them in this work.

Fig. 2 is a longitudinal section of the right forearm, hand, and third finger, from the same arm as fig. 1, and is viewed from the ulnar aspect. The radius is divided in its entire length; on its articular surface is the semilunar bone, and in front of it the os magnum and third metacarpal bone, the first phalanx, and a portion of the second, the third was not included in the section. The joints were not particularly prepared for the section. In the hand they are in the condition of partial extension, whilst the fingers are flexed from the effects of freezing. The skin is smooth on the dorsal aspect, whilst on the volar, which is rich in fat, it forms thick pads, giving rise to deep furrows. During extension these furrows appear as transverse lines, and do not correspond with the opposed articular surfaces of the joints. Those on the volar aspect of the root of the finger lie considerably further forwards than the corresponding metacarpo-phalangeal joints, and the subsequent furrow exceeds, though not to so great an extent, the joints between the first and second and third phalanges, consequently in disarticulation of a finger from the volar aspect the joint will not be opened if the knife be applied directly upon this furrow. The articulation will be far more certainly reached if the incision be made from the extensor aspect, after slightly flexing the finger, a little in front of the projection which the head of the bone makes with its distal phalanx. Corresponding with the more extensive expansion of the cartilage on the volar aspect, the cavity of the synovial membrane extends further upwards than on the extensor. The capsular ligament is moreover con-siderably strengthened by the tendinous expansions formed by the lateral ligaments, and which prevent too great an amount of extension of the finger. Immediately beneath the skin, and separated from it merely by bursal tissue, are the flexor tendons of the finger, of which the more superficial disappears at the first phalanx, and the tendon of the deep is shown passing on to its insertion into the ungual. These tendons are easily followed upwards, beneath the annular ligament to their muscles which form the chief bulk of the forearm.

The number of sections which could be made with advantage of the hand is unlimited, as in every change of position new and interesting forms arise. This is more particularly the case with regard to the region of the thumb, where section is especially suitable for the purpose of demonstrating the peculiar relations of the joint in dislocation. It is a pity that no more space can be afforded, and I must therefore refer the reader to Pirogoff's Atlas, fasc. iv B, tab. v and vi, where frontal sections of the hand are shown, and to fasc. iv A, tab. iv and v, which represents longitudinal sections of the thumb both in its normal and dislocated conditions.

Tab. XXVII.

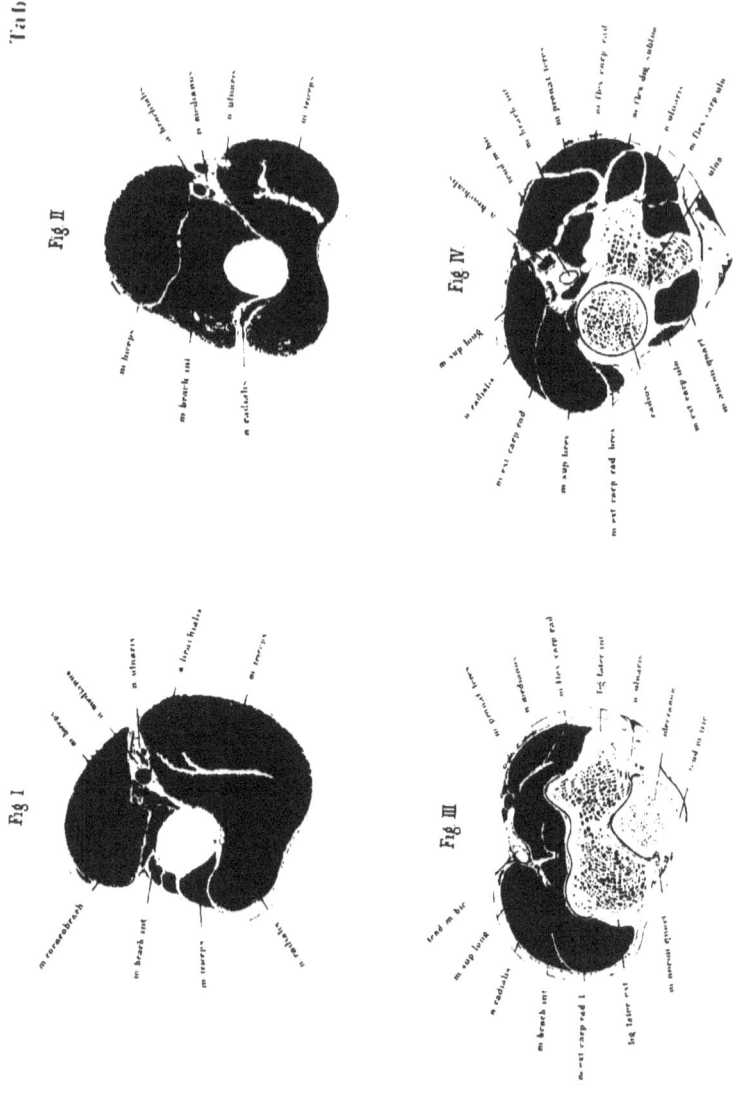

Fig I

Fig II

Fig III

Fig IV

PLATE XXVII

THE series of transverse sections from which the present and following plates were taken was made from the left arm of a man æt. 40. The arteries were injected. The forearm was slightly flexed and pronated. In order to obtain bearing points for the individual laminæ, a line was previously drawn passing through the middle of the biceps over the surface of the supinator longus to the thumb, and the uppermost points of each subsequent lamina lie in this line.

Fig. 1. In this instance the line of section passes through the middle of the arm, and its surface is seen from above downwards, hence we may imagine that we have the stump of an amputation of the right arm for examination, as has been before suggested in speaking of the lower extremity. The section is taken below the insertion of the deltoid, the biceps and triceps occupying the greater space. On the anterior aspect of the bone are portions of the brachialis anticus and coraco-brachialis; in the middle, to the right of the observer, and between the flexor and extensor muscles, are the great vessels and nerves, and the musculo-spiral nerve has already commenced its tortuous course accompanied by the superior profunda artery. This position of the nerve accounts for the fact that blows or injuries from behind are capable of compressing it so directly upon the humerus that paralysis may be the result. The separation of the muscular masses of the flexors and extensors is already at this level so decided, that the intermuscular aponeurosis appears in the frontal plane. The relation of the individual muscles is so clear as hardly to demand any particular explanation.

Fig. 2 is a section of the left arm in the middle of its lower third. The flexor and extensor muscles lie on both sides of the humerus, and the intermuscular aponeuroses are here still more clearly seen than in the preceding section. In the external intermuscular septum is the musculo-spiral nerve, which has nearly terminated its half turn round the humerus, and behind it is the origin of the supinator longus. On the inner side the ulnar nerve has already become distinct from the great vessels and mass of nerves. The brachial artery is on the inner border of the biceps, accompanied by its venæ comites, with the median nerve above it. Although its position is here very easily made out and its compression readily performed, there is great difficulty in isolating it and tying it unless the steps of the operation be carried out very correctly. The vessel, as the plate shows, cannot be directly cut down upon, as on account of nerves and veins which here often very freely anastomose, the operator may be much embarrassed; and experience has shown that the vessel may be easily missed; the surgeon must therefore make for the edge of the biceps, which is slightly in front of it, and open its sheath from the inner side, when he will come directly upon it.

The distance of the artery from the bone depends on the development of the brachialis anticus. In this instance, on account of the muscles in relation with it having become more developed, the vessel lies further from the bone than in the preceding section. Compare fig. 1 of this Plate, and also Plates X and XI, fig. 3.

Fig. 3. In this instance the plane of section passes through the lower end of the humerus and the olecranon. On the left side is the commencement of the capitellum with the end of the lateral epicondyle, on the left the trochlea with the middle epicondyle. The olecranon lies behind in the posterior supra-trochlear fossa. The extent of the cavity of the synovial membrane and capsule is indicated by a dark line.

Behind the olecranon is a large bursa between the skin and the tendon of the triceps. On the right, in the furrow between the olecranon and medial epicondyle is the ulnar nerve. To the left of the olecranon is the anconeus. The muscles of the arm are much reduced in bulk at their point of attachment. The origins, however, of the flexors and extensors of the hand and fingers, the pronator teres, and the supinator longus, the latter, on account of its high origin from the humerus, are more powerfully developed in the section. On the anterior aspect of the bones are masses of muscle, on the posterior merely ligaments and tendons, which allow of the bony

PLATE XXVII   185

prominences being clearly distinguished. This relation of the muscular masses, and the position of the vessel on the belly of the brachialis anticus, demonstrates the fact that all incisions which are intended to penetrate the joint should be arranged on its exterior aspect, as it can be here entered without fear of any considerable hæmorrhage, and the ulnar nerve alone requires care in looking after.

Fig. 4 is a section of the forearm through the head of the radius, which is clearly shown with the annular ligament, and the upper extremity of the ulna the lesser sigmoid notch of which lies in articulation with the radius. The brachialis anticus is now for the most part tendinous, and attached to the ulna on the other side of its tuberosity. The tendon of the biceps is behind the tuberosity, which lies below the surface of the section, and the bursa, between it and the upper part of this tuberosity, is indicated by a black line. The brachial artery lies in the middle in front of the joint, enclosed by the origin of the flexors and extensors. Its division into radial and ulnar is evident. In front of it is the communication between the superficial and deep veins, and shows why bleeding in this region is so copious, if contraction of the muscles around the deep vein be induced; it is, however, not possible to expose the intimate relations clearly by section. It may, however, be here explained that the "system" of the median vein does not only associate the trunks of the cephalic and basilic with each other, but also keeps up a communication with the deep veins accompanying the radial and ulnar arteries. The irregularly formed and generally small vein which lies in the bend of the elbow requires no particular note, as it possesses no further importance than the trunks which frequently approach close to the basilic and cephalic in the bend of the forearm. It is, however, worth while to designate this communication, which passes deep down, as median (it is named by Arnold, the deep median vein), and to denominate the oblique branches of communication between the basilic and cephalic as median basilic and median cephalic.

The mass of the flexor muscles is already at this level more strongly developed than in the preceding section. They predominate over the extensors, as will be still more clearly seen in the deeper section of the forearm.

24

PLATE XXVIII

FIG. 1. The section here passes through the upper third of the left forearm, and the ulna and radius exhibit surfaces of almost equal size, only the ulna with its sharp edge lies closer to the surface than the radius, which is embedded in muscle. The ulna can be readily felt throughout the entire length of the forearm, but the head and inferior extremity only of the radius. The edge of the ulna affords an easily distinguishable limit between the flexor and extensor muscles. The flexor carpi radialis forms the muscular limit on the flexor surface. It is placed with its tendinous border on the ulna, and covers over the deep flexor lying beneath it. On the opposite side of the ulna is the origin of the interosseous ligament, and in connection therewith fasciæ, which pass directly upwards, and consequently separate both groups of muscles. To the left lie the supinators and extensors, and to the right the pronator teres and flexors. Between both groups of muscles are seen the vessels, the ulnar artery deep down, with the interosseous springing from it, and above it the radial. One needs merely to divide the enveloping fasciæ, and to pull the muscle to one side to expose the radial artery. The deep position of the ulnar at this spot renders its ligature difficult. Of nerves the superficial branch of the radial is found below the supinator longus, the deep branch lying in the supinator brevis. The median is between the pronator teres and flexor sublimis digitorum, the ulnar between the latter and flexor carpi ulnaris.

Peculiar interest is attached to the supinator brevis, the function of which can be readily understood by reference to this section. Passing outwards from the ulna (its upper set of fibres from the epicondyle are not seen), it wraps round the radius so that it must by its contraction roll it outwards. The space between it and the radius is taken up by the tendon of the biceps, which from the nature of its attachment assists in supination.

Fig. III — Fig. I — Fig. IV — Fig. II

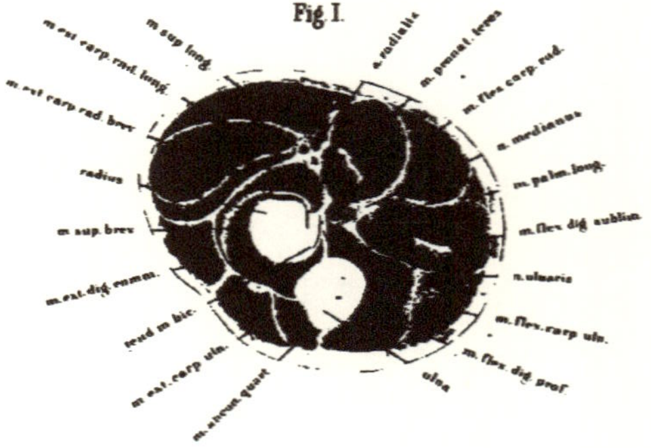

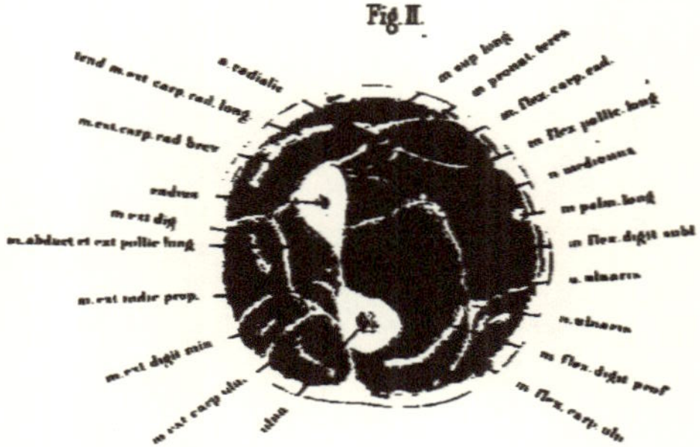

Tab. XXVIII.

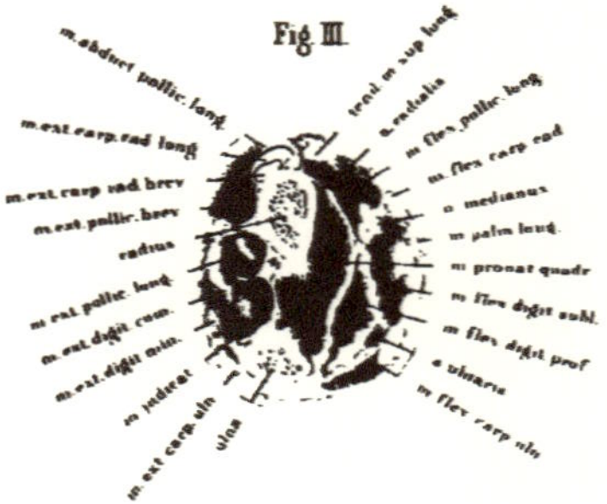

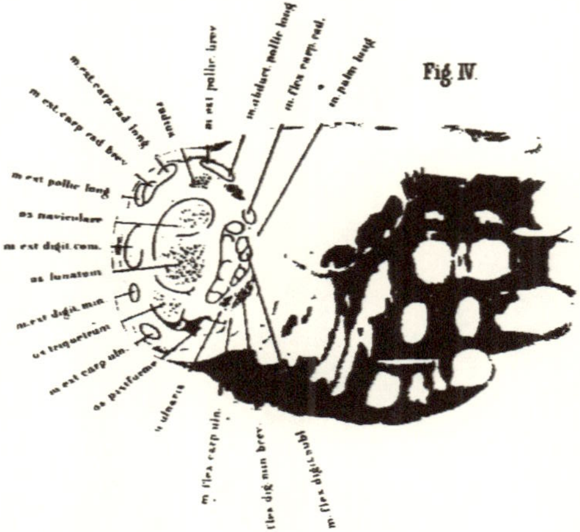

PLATE XXVIII   187

Fig. 2. In this plate, which shows a section through the middle of the left forearm, there is considerably greater difficulty in recognising the relations of the individual structures than in the preceding, and this difficulty is not so much from the number of muscles, but from the absence of the fascial septa which limit the individual groups. The interosseous ligament alone forms with the skeleton an absolute limit, and this does not extend throughout the entire breadth of the section. The ulna and radius present their sharp edges to each other, and are bound together by the interosseous ligament; on the right is the mass of the flexors, and on the left that of the extensors. Both groups of muscles are separated from the radius by a very thin fascial covering, which is attached to the radius and it encloses the radial artery and veins. This vessel is at this level easily found beneath the inner border of the supinator longus. The ulnar artery and nerves are here nearer the surface than in the preceding section, and the surgeon has only to make an incision between the flexor carpi ulnaris and the flexor sublimis digitorum to reach it. The fascial lamina passing from it to the median nerve, and which is prolonged beneath the origin of the pronator teres to the radius, divides the deep layer consisting of the flexor profundus digitorum and flexor longus pollicis from the superficial flexors, in which the flexor sublimis digitorum has penetrated beneath the palmaris longus which has already become tendinous. In like manner, on the opposite side of the interosseous ligament the extensors are divided into a superficial and deep layer; and the extensor ossis metacarpi pollicis and the extensor secundi internodii are already shown.

If the flexor surface be compared with the extensor with reference to the mass of its muscles, it is seen at once that the flexor considerably predominates over the extensor, and farther that the main trunks of the vessels lie on the flexor surface. If the surgeon has a choice of flap in amputation of the forearm in this region, provided there is nothing to the contrary, he should form his flap chiefly from the flexor surface, as much on account of the quantity of soft parts as for the nourishment afforded the stump by the vessels. The formation of flaps from the extensor aspect is much more difficult of performance on account of the closeness of the bones to the surface.

Fig. 3 is a section of the left forearm in its lower third. The radius has become considerably thicker in section. Its surface is covered by the broad pronator quadratus, which from its attachment to the ulna, rolls the radius over to the position of pronation. Beneath it is the interosseous ligament, and on both sides of it the interosseous vessels. The proximity of the radio-carpal joint is evident from the presence of the tendons of the muscles. The flexors and extensors even to the flexor carpi ulnaris have become tendinous, consequently the radial artery lies free, covered only by skin and fascia, hence affording the readiest means of feeling the pulse, and is here very easily ligatured. The ulnar artery, on the contrary, is still covered over by the tendinous border of the flexor carpi ulnaris, which must be drawn aside in order to reach it. On the extensor aspect are the long muscles of the thumb, and passing upwards from below the radial extensors of the carpus. At their points of crossing, bursæ are developed to prevent their rubbing against one another. Over-use of these muscles, such as with mowers, may cause inflammation of these bursæ and form a tumour over this locality (teno-synovitis). The number of muscles again on the flexor surface exceeds that of the extensor surface in this section. The mass of the muscles has, however, so much diminished that in amputation of the forearm in this region the flap à la manchette is preferable, as the plate sufficiently explains.

Fig. 4. In this plate the section passes through the carpus. The tendons only of the muscles are now shown with their bursal tissue, the presence of which is indicated by the numerous black lines around the divided tendons, the only muscular tissue cut being that of the ball of the little finger. Only a small portion of the radius, the root of its styloid process, is shown.

The bones of the carpus seen, are the semilunar, scaphoid, and cuneiform, and the articulation between it and the pisiform, the surface of which is seen anteriorly, has been opened.

The three bones of the first row represent a surface, the individual portions of which are moveable, and which articulates with the radius, and with the ulna by means of the inter-articular fibro-cartilage. The

PLATE XXVIII   189

section of these three bones, as here represented, has not the form of an ellipse, but resembles a parallelogram, with its angles rounded off. The articular surfaces of these bones approximate to the spherical form, being received into an oval hollow, somewhat in the same manner as the head of the humerus into the glenoid cavity of the scapula.

PLATES XXIX (A, B) AND XXX

THE body from which this preparation was made was quite recent, twenty-five years of age, in the last month of pregnancy, and received in the condition of rigor mortis. Beyond the constriction of the neck produced by the means of death (hanging) no abnormality existed. The condition of the genitals corresponded with an advanced stage of pregnancy, and were injected and succulent. The method of preparation was carried out in the usual manner. The foetus, which was divided in the section of the body, was subsequently restored to its original condition, so as to afford a representation of its former position in the uterus. I chiselled out the foetus and the liquor amnii from the left side of the body, and moistened the surface of the section of the uterus, and then froze it on the right side. The portion now lying in the right half of the uterus remained then for the purposes of representation as an untouched foetus. The left half of the uterus and its appendages, after the removal of the rest of the liquor amnii, was represented as empty. The foetus, which was in the second position of the head, was a well-formed female. The vulva were closed and the nails well developed. Its entire length was about twenty-three inches, its weight without the cord about six pounds. The cord was divided, and passed to the placenta between the head and right arm, the placenta being placed downwards and on the right side of the uterus. The child, as the plate shows, lay mostly in the right half of the uterus. In the section more than the right half of the head which was sawn obliquely, was removed. Moreover, the left arm and a portion of the right shoulder were divided longitudinally, and the forearm being placed at right angles with it, transversely, as well as a portion of the right leg, which extended towards

Tab. XXIX A.

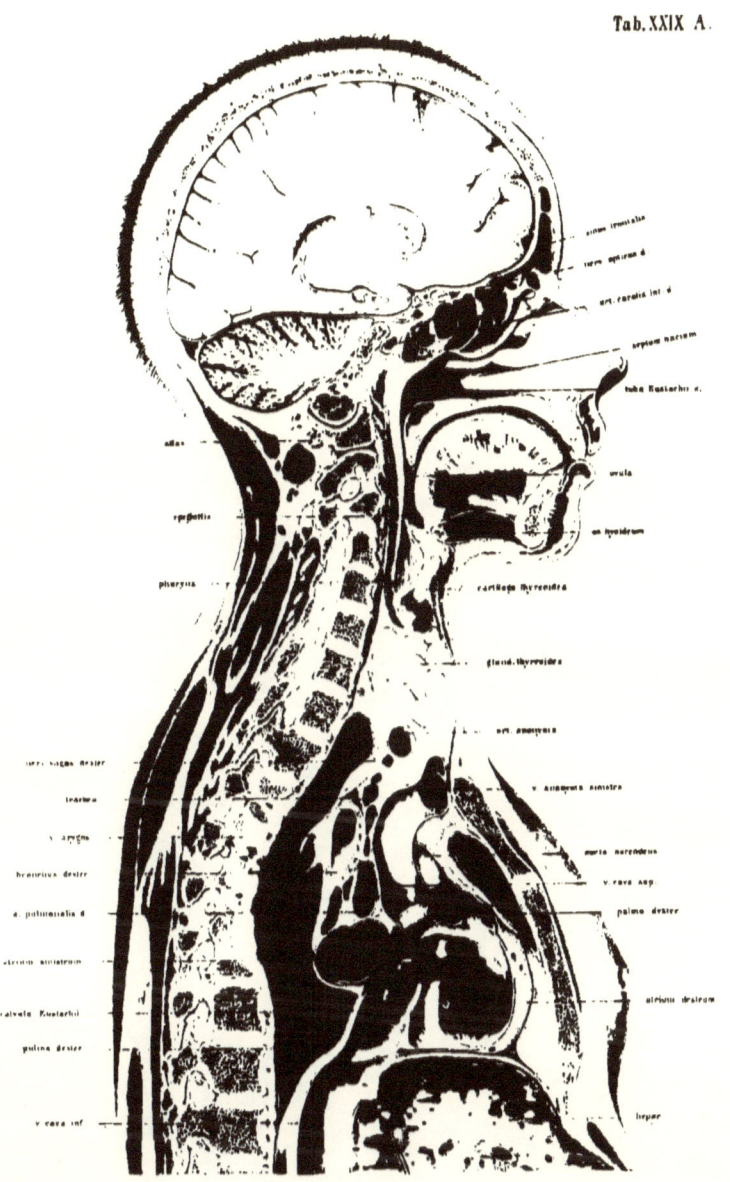

sinus frontalis

turs. optieus d.

art. carotis int d.

septum narium

tuba Eustachii d.

uvula

os hyoideum

cartilago thyreoidea

gland. thyreoidea

art. anonyma

v. anonyma sinistra

aorta ascendens

v. cava sup.

pulmo dexter

atrium dextrum

hepar

nerv. vagus dexter

trachea

v. azygos

bronchus dexter

a. pulmonalis d

atrium sinistrum

valvula Eustachii

pulmo dexter

v. cava inf.

atlas

epiglottis

pharynx

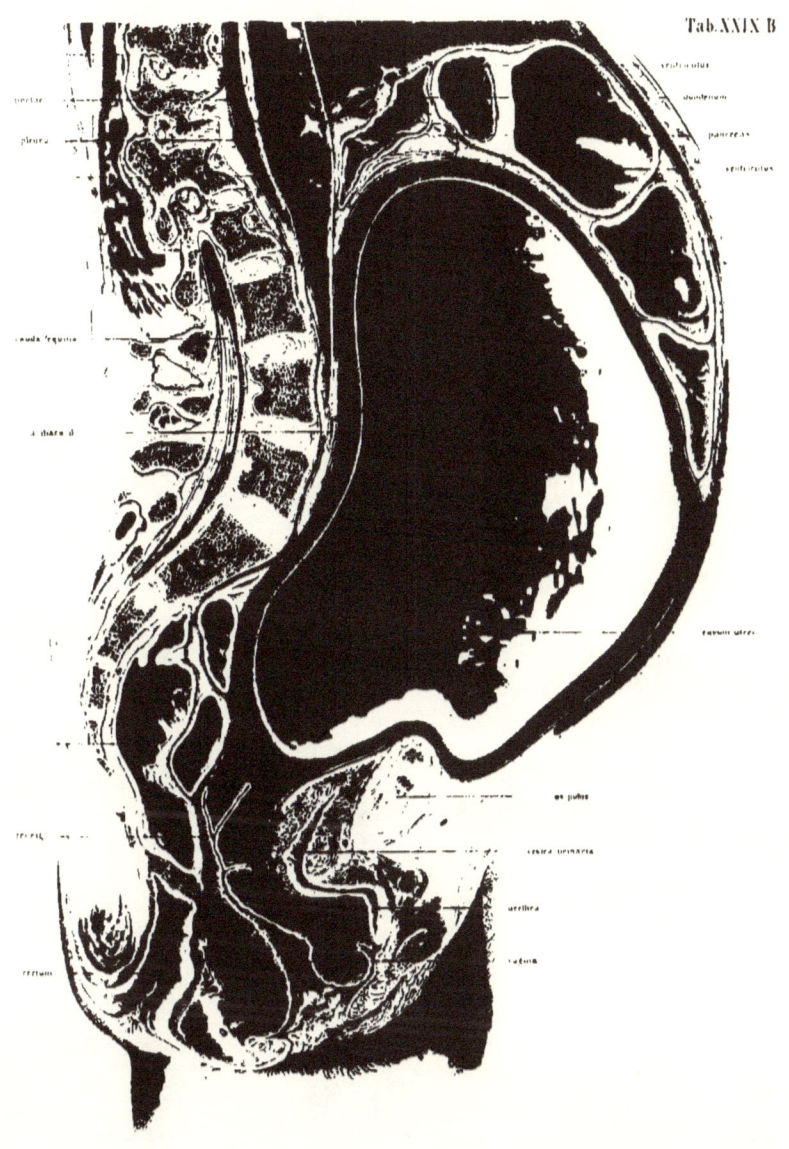

Tab.XXIX B

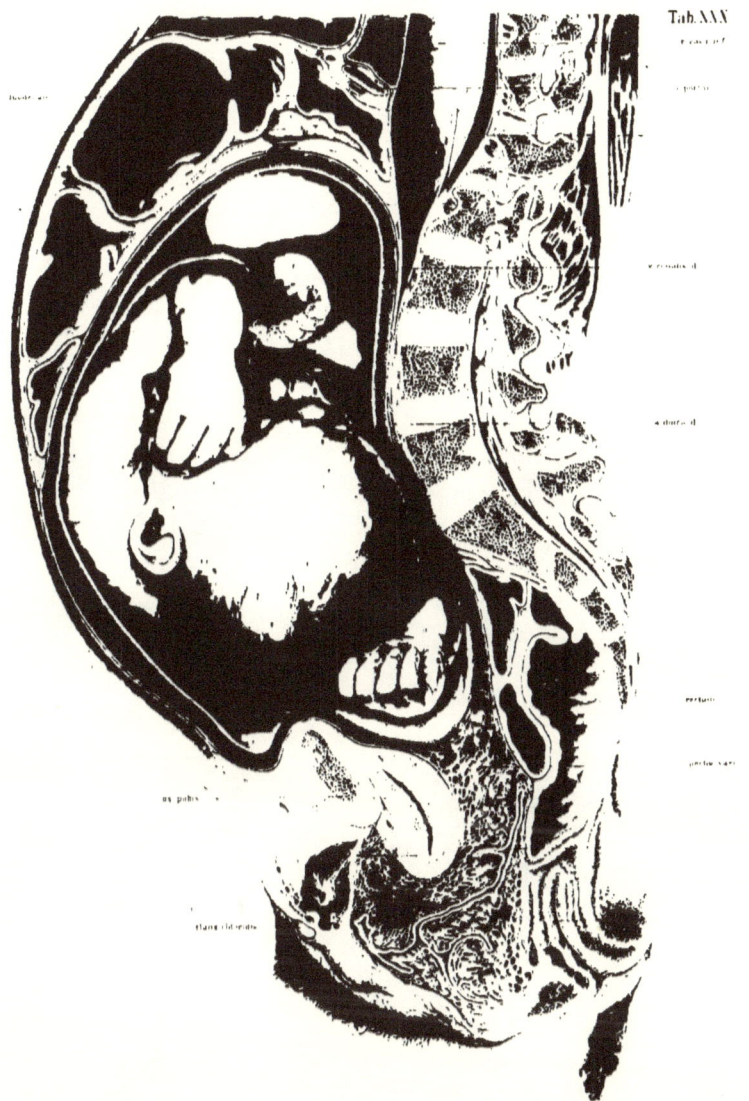

Tab. XXX

tho left side. The left knee was moreover grazed by the saw. The back and belly lay in the right half of the uterus, and the greater portion of the liquor amnii remained in the left. As the relations of this oblique section of the fœtus offer points of no peculiar interest, I have refrained from reproducing the corresponding plate of the large atlas in this small edition.

The uterus is so folded over the symphysis that its anterior wall forms a kind of sac, indicating a condition of relaxation. The numerous large veins in its tissue are shown in the plate in the wall as simple strokes, their lumina becoming recognisable only when their walls were separated from each other; they appear patent, however, in tho vaginal portion of tho uterus and in the vagina itself. Tho vaginal portion of the uterus is proportionately deep, and for the most part lies in the left half of the body, the section having passed through its right half and opened merely the first portion of the cervix, as shown in Plate XXIX B. It was filled with viscid mucus and opened into tho cavity of the uterus, about one fifth of an inch below the plane of section, so that its upper half could not be seen. The length of the vagina at this period of pregnancy makes it probable that the woman was not a primipara, notwithstanding that there were no cicatrices on the abdominal parietes, and the os internum was so narrow that only a very small sound could pass it. The number of veins met with in the right half of the vagina and their swollen condition is remarkable, and their lumina are peculiarly well seen in the left half of the preparation, Plate XXIX B. The falling in of the vaginal portion of the uterus is remarkable, considering the empty contracted condition of tho bladder. The latter has slipped down bodily from the inner surface of tho symphysis, and is so completely displaced that the course of tho urethra has become bent at an angle. The external os lies in the hollow of the under border of the symphysis, although, according to Moreau, it corresponds at the end of pregnancy with the level of the upper border of the symphysis, and is still higher according to Schultze.

Tho level of the fundus corresponds nearly with the under border of tho first lumbar vertebra; a more accurate definition cannot be given, as the highest point of the uterus was not included in the section, as it inclined more to the right side. This is almost the level given by Moreau, and according to the measurements of Schultze ('Wandtafeln,' taf. vi), it would appear to be the second lumbar vertebra. As the parts in the meanwhile began to thaw, a more accurate measurement in this particular could not be made.

The depth of the cavity of the uterus and its connections, and of the entire cavity of the abdomen, is less than is usually admitted. Notwithstanding the size of the fœtus it is not improbable that the attitude of the body had some influence in this respect, and that lying horizontally on the back the uterus obtained a kind of fulcrum on the vertebral column, whilst in the upright position the yielding walls of the abdomen are pushed forwards. It is farther to be remembered that in dead bodies generally in consequence of the high position of the diaphragm, the depth of the cavity of the abdomen is less than during life.

In the present instance the distance of the lumbar vertebræ from the anterior wall of the abdomen was almost one third of the entire sagittal diameter of the body at its point of greatest distension; whilst in the body which in Plate II is represented in the second month of pregnancy, the lumbar spine projects slightly beyond the middle of this diameter.

Finally, the vessels were in this case uninjected—a circumstance which is to be taken into consideration in estimating the thickness of the walls of the uterus.

The cavity of the abdomen extended tolerably far up in comparison with its slight depth. The highest point of the diaphragm reached the level of the seventh dorsal vertebra, whilst in males, and unimpregnated females of middle age it would extend only as far as the ninth or tenth.

The mass of the intestines was pushed downwards, and chiefly lodged in the left upper half of the abdominal cavity. The pyloric extremity of the stomach was bent at an acute angle backwards and to the left side, so that it was twice cut through. The upper horizontal portion of the duodenum was directed backwards. On the left half of the body the duodenum is contracted against the pylorus, and moreover shows the opening of the pancreatic and choledic ducts. Below the duodenum is the pancreas. The liver and spleen are not enlarged. The latter measured 5·5 inches long, 3·8 inches broad, and 2 inches thick.

The duodenum and pyloric portion of the stomach had pushed the fundus uteri, which lay more to the right, together with the rest of the intestines over to the left side.

The rectum, which was tolerably distended, bent round to its iliac flexure, towards the right side whilst yet in the pelvic cavity, so that this flexure is met with in the course of the section. Between the rectum—the folds of which on the right side are so disposed that they might be taken for the valves of Kerkring in the plate—and the uterus, is a coil of small intestine, the lowest portion of the ileum passing to the ascending colon, a disposition which does not usually happen with the impregnated uterus, but only in anteversion.

In examining the limit of the peritoneum in the pelvis, it must not be confounded with the fascia, represented rather too thick, which passes down between the uterus, bladder, and rectum. The peritoneum is applied for only a short extent to the posterior wall of the vagina, and envelopes nearly half of the posterior wall of the contracted bladder, whilst the fasciæ, which enclose a loose, lax cellular tissue, pass forwards nearly to the internal orifice of the urethra, and posteriorly close to the end of the rectum.

The thoracic cavity appears shallow, in consequence of the high position of the diaphragm, but, on the other hand, very wide in the antero-posterior diameter, as may be seen by comparing this preparation with the section of the female subject in Plate II. But on the strength of this, an enlargement of the base of the thorax during pregnancy is not necessarily to be inferred, as measurements for comparison are wanting before and after it. Although it may appear plausible to explain the unvarying size of the spirometer during pregnancy, by the fact that the diminution of the thoracic space dependent on the rising of the diaphragm is compensated for by the traction of the abdominal muscles acting over the uterus like a pulley, the anatomical relations in this respect are not yet determined. Gerhard found, by measurements on living bodies, that in forty-two females in advanced pregnancy the diaphragm was in thirty-six cases in a normal position, in five it was deeper,

and only in one higher. Dorn in his measurements by means of the cyrto-
meter on living females in advanced pregnancy and in lying-in women,
found that in most cases the bases of the thorax had a greater breadth
during pregnancy than after delivery, but, on the other hand, its depth was
less from before backwards. When the uterus was empty this relation was
reversed, the thorax collapsed on both sides, the transverse diameter
decreased, and the vertical diameter increased ('Bericht über die Natur-
forchenversammlung zu Giessen,' 1865, p. 225).

In the cavity of the thorax, in consequence of the scoliosis of the spine,
the section deviated to the right of the middle line, so that traversing the
lumina of the superior and inferior venæ cavæ, the right auricle and root of
lung are met with, consequently the relation of the openings of both veins
into the heart are clearly seen. The inferior vena cava, which receives the
hepatic vein just before its entrance into the heart, comes from behind into
the right auricle, whilst the superior vena cava opens considerably further
forwards. The axes, therefore, of the cavæ form an angle, which, owing to
the convexity of the septum auriculorum, is rounded off. The eminence,
behind which lies the left auricle, is the tuberculum Loweri. On the right
half of the body is noticed the external wall of the right auricle, whilst on
the left a view is obtained of the left ventricle, in front of the entrance to
which is still a small portion of the rudimentary Eustachian valve. This
valve limits the posterior portion of the right auricle, in which is still to be
seen the original position of the foramen ovale. In the anterior portion of
the auricle, to which the superior cava tends, the bulbus aortæ forms a
flattish protuberance. The aorta itself is not seen entirely, a portion of it
only being exposed. It rises in front of the superior vena cava, and then
disappears below the left innominate vein.

The section of the lung seen in Plate XXIX is that of the right.

The soft parts of the neck are considerably dislocated towards the left
side, owing to the hypertrophied thyroid body. The trachea lies so far
over to the left side that only a small portion of the thyroid cartilage is met
with.

The brain was divided through its right half, the radiation of the fibres
of the right corpus callosum being thus shown. Beneath it is the

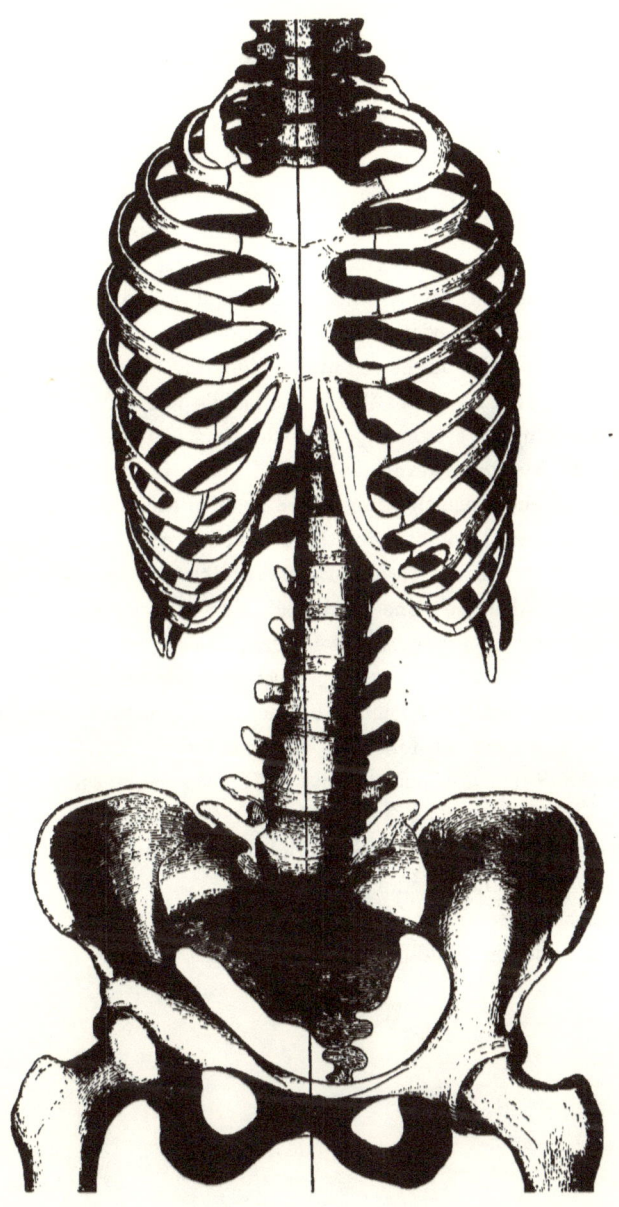

descending cornu of the right lateral ventricle with the pes hippocampi. Beneath the dura mater, in the right half of the preparation, a portion of the Gasserian ganglion and some fibres of the fifth nerve are seen.

The relations of the skeleton, however, are of the greatest importance. I had therefore, after all the plates were drawn, the halves of the skeleton macerated, and the parts as accurately as possible adjusted with regard to each other, as represented in the adjoining woodcut. It presents a slightly scoliosed pelvis, with a like condition of the spine. It shows moreover that the deviation of the line of section from the middle line was not so considerable as the plate might suggest. The section passed through the pelvis, as near as possible in the middle line, externally and to the right of the lumbar vertebræ, meeting the dorsal at their articulation with the ribs, and passing again in the cervical region to the middle of the spinal column, and subsequently again to the right in the skull.

Beyond the scoliosed condition of the spine there was nothing worthy of remark, except that there were two cervical ribs, one complete on the right side, and a rudimentary one on the left side of the seventh cervical vertebra. There were seven cervical vertebræ, but only eleven dorsal and five lumbar. There was a rudimentary process from the fifth lumbar which was attached to the upper portion of the sacrum. The measurements of the pelvis in inches were as follows :—The conjugata vera 3·8 in. (the conjugata at the narrowest points being 3·7) ; the right sacro-cotyloid 2·8 in. ; the left sacro-cotyloid 3·2 in. ; the transverse diameter 5·8 in. ; the left oblique diameter, 5·08 in., and the right oblique diameter 5·6 inches. The sacrum was 4·5 in. deep and 4·8 in. broad.

The question arises whether, in a weak obliquely contracted pelvis, showing such a variation, child-birth be possible without surgical aid.

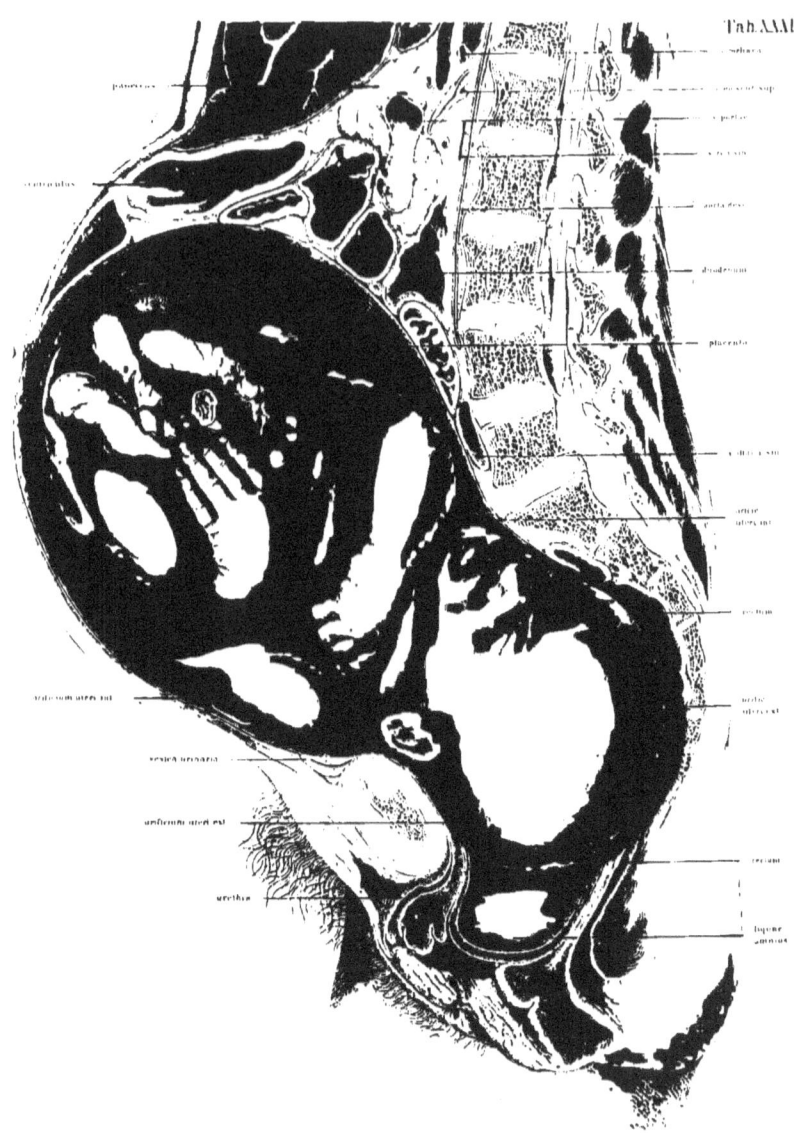

Tab.XXXI

PLATE XXXI*

The body from which this plate was made, was that of a person thirty-five years of age, who died from drink at the commencement of labour. An examination of the genitals showed that the liquor amnii had not escaped. After having been prepared in the usual way, a section was made in the mesial plane from below upwards. The symphysis was not however exactly divided at its centre, but the deviation was so slight that it need not be regarded.

After the drawings were made of the right half of the body, and completely finished, the maternal structures were removed, in order to obtain the other half of the child uninjured and in its original position.

The child was a well-formed male of about six pounds weight including the cord which passed downwards under the left leg, whence it was bent upwards and lay over the left ankle joint, being reflected sharply on to the placenta which was attached to the upper portion of the uterus. The cord must have been cut through on removing the left half of the child, as I afterwards found its placental insertion in the left half of the body. I had divided it close to its placental extremity, and it was so firmly pressed against the child, that it could be with difficulty removed without inducing a change in the position of the left lower extremity.

The child's head as is seen in the plate is apparently in the second position, and was on the point of being born at the death of the mother. The natural rotation of the head in the pelvis has commenced, being

* As this chapter refers almost entirely to the section of the child, and the corresponding plates are not reproduced in this small edition, I have thought it advisable to omit such portions of it as are not illustrated directly to the accompanying plate, and to advise the reader interested in the matter to consult Prof. Braune's 'Die Lage des Uterus und Fœtus am Ende der Schwangerschaft,' which has been already translated into English.—Tr.

turned more to its right side than its trunk. The shoulders are entirely in the false, whilst the head has already entered the true pelvis. The propulsive force must have been considerable, as the fœtal head is large and the pelvis not particularly wide, and the evidences of this force are shown from the form of the head. Its posterior portion is pointed or pearshaped, and an examination showed there was considerable effusion of blood on the skull.

Further it appears that owing to the strong contractions of the uterus upon the child the joints nowhere exhibit their rounded form with the freely flexed position of the extremities, as it is packed in the smallest space possible.

The skin was thrown into sharp ridges. The nape of the neck appeared as a narrow chink between the deeply folded skin of the back and occiput. The bulging out of the head in the region of the left ear is remarkable, being produced by the pressure exerted by the narrow pelvis. Just below it is the section of the symphysis. A deep notch is produced in the left arm by the internal os, which has also left traces behind on the right forearm which passed down longitudinally over it.

The uterus itself is of especial interest, the relations of which are well seen after the removal of the entire child, and a particular plate of its empty cavity seemed necessary. The uterus held the child firmly, and had no folds in it, from which a lax condition of its walls might be inferred. It was placed with its long axis directed nearly vertically towards the plane of the inlet of the pelvis, so that it had the appearance of having remained in a state of contraction after death. The internal os lay somewhat over the inlet, and is noticeable from the lumen of a large vein with a smaller one beside it, the only veins which were found patent in the walls of the uterus, in consequence of the blood remaining in them.

In the empty cavity of the uterus, the internal os appeared as a freely projecting semicircle, an inch and a half above the symphysis, and four fifths of an inch above the premontory of the sacrum. The external os, which was completely dilated, appeared as a small projection. It was drawn out obliquely, from the region of the lower border of the symphysis

PLATE XXXI 199

towards the articulation between the sacrum and coccyx, and surrounded the protruding occiput of the fœtus. The opening of the right Fallopian tube was well seen in the upper third of the uterus.

The depth of the uterine cavity from the horizontal plane of the surface of the section to its lowest point was 2·6 inches, and the distance of the internal os to the fundus 6·6 inches. The distance of the external os from the inner in the axis of the pelvis was 4·5 inches. The thickness of the walls of the uterus varied considerably in different places.

The placenta lay in the floor of the uterus, and chiefly in its left half, therefore the origin of the cord must have been divided in the section. Although the patency of the rectum was here and there retained, the bladder was empty and contracted. Behind the symphysis its walls had become so thin as to be hardly recognisable. Above and below the symphysis, where the pressure had not been so great, it was thicker, and consequently better seen. On filling the bladder it became distended upwards, so that the anterior walls of the abdomen must have been lifted up slightly from the uterus.

The protrusion of the abdominal parietes by the uterus is here more considerable than in the preceding case, in which that organ occupies a different position, although not so marked as would appear at first sight. In both the spine projects beyond a third of the entire depth of the trunk.

Unfortunately, during the removal of the left half of the mother in order to obtain the necessary view of the child, the skeleton was destroyed, so that it was impossible to restore it as was done in the previous case.

As the vertebræ had no lateral deviations so the section passed exactly in the mesial line. The kyphotic curvature of the thoracic and cervical portions, though marked, is of no particular interest as regards the relations here shown.

The relations of the intestines and stomach, the former being, as in the other body, pushed upwards, afford nothing worthy of remark. The liver and spleen were normal. The former weighed about three pounds, and displayed a protuberance of its anterior wall, as in the first instance, the latter weighed about seven ounces, and measured 5·6 inches

long, 2·8 inches broad, and 1·2 inches deep, was therefore rather less than
generally met with in advanced pregnancy.

The depth of the cavity of the abdomen, measured from the symphysis
to the cupola of the diaphragm, is pretty much the same as in the former
case, whilst there is considerable difference in that of the thorax. I need
call no particular attention in this case to the diameter of the base of the
thorax, which naturally exceeds that of the female at the second month
represented at Plate II. Any decided measurement as to the depth of the
thorax during pregnancy can only be made on the living body, when the
relations of the chest cavity before and after delivery must be determined.

The relations of the heart, trachea, larynx, mouth, and brain were
entirely normal.